JUNGLE FOLK

INDIAN NATURAL HISTORY
SKETCHES BY DOUGLAS DEWAR

I0119934

JUNGLE FOLK

PREFACE

IT is not of the bigger jungle folk that I write—of lions, tigers, leopards, bears, bison, or even deer and antelopes ; for of these it is vouchsafed to no man— not even to the *shikari*, who spends years in the jungle —to obtain more than an occasional fleeting glimpse.

The subjects of my theme are the lesser fry—vivacious mynas, noisy babblers, vociferous cuckoos, silent herons, beautiful pittas, graceful wagtails, elegant terns, melodious rock-chats, cheeky squirrels.

A cheery crowd are these. The man who passes his days in India without knowing them misses much of the pleasure of life.

D. DEWAR.

CONTENTS

JUNGLE FOLK

JUNGLE FOLK

I

OF INDIAN BIRDS IN GENERAL

LITERARY critics seem to be agreed that we who write about Indian birds form a definite school. " Phil Robinson," they say, " furnished, thirty years ago, a charming model which all who have followed him in writing seem compelled to copy more or less closely." Mr. W. H. Hudson remarks : " We grow used to look for funny books about animals from India, just as we look for sentimental natural history books from America."

In a sense this criticism is well founded. Popular books on Indian ornithology resemble one another in that a ripple of humour runs through each. But the critics err when they attempt to explain this similarity by asserting that Anglo-Indian writers model themselves, consciously or unconsciously, on Phil Robinson, or that they imitate one another. The mistake made by the critics is excusable. When each successive writer discourses in the same peculiar style the obvious inference is that the later ones are guilty of more or less

unconscious plagiarism. The majority of literary critics
in England have not enjoyed the advantage of meeting
our Indian birds in the flesh. To those who do possess
this advantage it is clear that the Indian birds themselves are responsible for our writings being " funny."
We naturalists merely describe what we see. The avifauna of every country has a character of its own. Mr.
John Burroughs has remarked that American birds as a
whole are more gentle, more insipid than the feathered
folk in the British Isles. 'Still greater is the contrast
between English and Indian birds. The latter are to
the former as wine is to water.

India is peculiarly rich in birds of character. It is
the happy hunting-ground of that unique fowl, *Corvus
splendens*—the splendid crow—splendid in sagacity,
resource, adaptiveness, boldness, cunning, and de-
pravity ; a Machiavelli, a Shakespeare among birds, a
super-bird. The king crow (*Dicrurus ater*) is another
superlative. He is the Black Prince of the bird kingdom. He is the embodiment of pluck. The thing in
feathers of which he is afraid has yet to be evolved.
Like the mediæval knight, he goes about seeking those
upon whom he can perform some small feat of arms.

When we turn to consider the more outward characteristics of birds, the peacock, the monal pheasant,
the " blue jay," the oriole, the white-breasted king-
fisher, the sunbird, the little green bee-eater, and a host
of others rise up before us. Of these some, showily
resplendent, compel attention and admiration ; others
of quieter hues possess a beauty which cannot be
appreciated unless they be held in the hand and each

feather minutely examined. At the other extreme stands the superlative of hideousness, the ugliest bird in the world—*Neophron ginginianus*, the scavenger vulture. The bill, the naked face, and the legs of this creature are a sickly yellow. Its plumage is dirty white, with the exception of the ends of the wing feathers, which are shabby black. Its shape is displeasing to the eye, and its gait is an ungainly waddle. Yet even this fowl looks almost beautiful as it sails on outstretched pinions, high in the heaven. Between the extremely beautiful and the extremely ugly birds we meet with another class of superlatives—the extremely grotesque. This class is well represented in India. The great hornbill—*Dichoceros bicornis*—and the adjutant—*Leptoptilus dubius*—are birds which would take prizes in any exhibition of oddities. The former is nearly four and a half feet in length. The body is only fourteen inches long, being an insignificant part of the bird, a mere connecting link between the massive beak and the great, loosely inserted tail. The beak is nearly a foot in length, and is rendered more conspicuous than it would otherwise be by a structure known as the casque. This is a horny excrescence, nearly as large as the bill, which causes the bird to look as though it were wearing a hat which it had placed for a joke on its beak rather than on its head. The eye is red, and the upper lid is fringed with eyelashes, which add still further to the oddity of the bird's appearance. The creature has an antediluvian air, and one feels when contemplating it that its proper companions are the monsters that lived in prehistoric times. The actions of the hornbill are in keeping with its appearance. It is the clown of the forest.

Even more grotesque is the adjutant. This is a stork with an enormous bill, a tiny head, and long neck, both innocent of feathers. From the front of the neck hangs a considerable pouch, which the bird can inflate at will. Round the base of the neck is a ruff of white feathers that causes the bird to look as though it had donned a lady's feather boa. It is the habit of the adjutant to stand with its head buried in its shoulders, so that, when looked at from behind, it resembles a hunch-backed, shrivelled-up old man, wearing a grey swallow-tailed coat. It looks still more ludicrous when it varies the monotony of life by kneeling down; its long shanks are then stretched out before it, giving the idea that they have been mistakenly inserted hind part foremost ! Its movements partake of the nature of a cake-walk. " For grotesque devilry of dancing," writes Lockwood Kipling, " the Indian adjutant beats creation. Don Quixote or Malvolio were not half so solemn or mincing, and yet there is an abandonment and lightness of step, a wild lift in each solemn prance, which are almost demoniacal. If it were possible for the most angular, tall, and demure of elderly maiden ladies to take a great deal too much champagne, and then to give a lesson in ballet dancing, with occasional pauses of acute sobriety, perhaps some faint idea might be conveyed of the peculiar quality of the adjutant's movements." If the hornbill be the clown of the forest, the adjutant is the buffoon of the open plain.

Consider for a little avine craftsmanship, and you

will find no lack of superlatives among our Indian birds. The weaver-bird (*Ploceus baya*), the wren-warbler (*Prinia inornata*) are past masters of the art of weaving. The tailor-bird (*Orthotomus sutorius*), as its name implies, has brought the sartorial art to a pitch of perfection which is not likely to be excelled by any creature who has no needle other than its beak.

If there be any characteristic in which Indian birds are not pre-eminent it is perhaps the art of singing. A notion is abroad that Indian birds cannot sing. They are able to scream, croak, and make all manner of weird noises, but to sing they know not how. This idea perhaps derives its origin from Charles Kingsley, who wrote : " True melody, it must be remembered, is unknown, at least at present, in the tropics and peculiar to the races of those temperate climes into which the song-birds come in spring." This is, of course, absurd. Song-birds are numerous in India. They do not make the same impression upon us as do our English birds because their song has not those associations which render dear to us the melody of birds in the homeland. Further, there is nothing in India which corresponds to the English spring, when the passion of the earth is at its highest, because there is in that country no sad and dismal winter-time, when life is sluggish and feeble. The excessive joy, the rapture, the ecstasy with which we greet the spring in the British Isles is, to a certain extent, a reaction. There suddenly rushes in upon the songless winter a mighty chorus, a tumult of birds to which we can scarcely fail to attach a fictitious value. India possesses some song-birds which can hold their

own in any company. If the shama, the magpie-robin,
the fan-tailed fly-catcher, the white-eye, the purple
sunbird, the orange-headed ground thrush, and the
bhimraj visited England in the summer, they would
soon supplant in popular favour some of our British
song-birds.

Another feature of the Indian avifauna is its richness
in species. Oates and Blanford describe over sixteen
hundred of these. To the non-ornithological reader
this may not convey much. He will probably obtain
a better idea of the wealth of the Indian avifauna when
he hears that among Indian birds there are numbered
108 different kinds of warbler, 56 woodpeckers, 30
cuckoos, 28 starlings, 17 butcher-birds, 16 kingfishers,
and 8 crows. The wealth of the fauna is partly accounted for by the fact that India lies in two of the
great divisions of the ornithological world. The
Himalayas form part of the Palæarctic region, while
the plains are included in the Oriental region.

Finally, Indian birds generally are characterised by
their fearlessness of man. It is therefore comparatively
easy to study their habits. I can count no fewer than
twenty different species which, during past nesting
seasons, have elected to share with me the bungalow
that I happened to occupy. Is it then surprising that
an unbounded enthusiasm should pervade the writings
of all Indian naturalists, that these should constantly
bubble over with humour ? The materials on which
we work are superior to those vouchsafed to the
ornithologists of other countries. Our writings must,
therefore, other things being equal, excel theirs.

II

RESPECTABLE CUCKOOS

THE general public derives its ideas regarding the manners and customs of the cuckoos from those of *Cuculus canorus*, the only species that patronises the British Isles. " The Man in the Street," that unfortunate individual who seems never by any chance to catch hold of the right end of the stick, is much surprised, or is expected to express great surprise, when he is informed that some cuckoos are not parasitic, that not a few of them refuse to commit their eggs and young ones to the tender mercies of strangers. The non-parasitic cuckoos build nests, lay eggs and sit on them, as every self-respecting bird should do. All the American species of cuckoo lead virtuous lives in this respect. But the Western Hemisphere has its evil-living birds, for many cow-birds—near relatives of the starlings—lay their eggs in the nests of their fellow-creatures ; some of them go so far as to victimise the more respectable members of their own brotherhood.

There are several upright cuckoos among our Indian birds, so that there is no necessity for us to go to America in order to study the ways of the non-parasitic species of cuckoo.

First and foremost among these is our familiar friend the coucal, or crow-pheasant (*Centropus sinensis*), known also as the lark-heeled cuckoo, because the hindmost of its toes has a long straight claw, like that of the lark. This cuckoo is sometimes dubbed the " Griff's pheasant," because the new arrival in India frequently mistakes the bird for a pheasant, and thereby becomes the laughing-stock of the " Koi-Hais."

It always seems to me that it is not quite fair to poke fun at one who makes this mistake. A man cannot be expected to know by instinct which birds are pheasants and which are not. The coucal is nearly as large as some species of pheasant, and rejoices in a tail fully ten inches long ; moreover, the bird is usually seen walking on the ground. Further, Dr. Blanford states that crow-pheasants are regarded as a great delicacy by Indian Mohammedans, and by some Hindu castes. I have never partaken of the flesh of the coucal, and those Europeans who have done so do not seem anxious to repeat the experiment. Its breast is smaller than that of a *dak* bungalow *murghi*, for its wing muscles are very small. As to its flavour, Col. Cunningham informs us, in his volume *Some Indian Friends and Acquaintances*, that " a young fellow, who had recently arrived in the country, complained with good reason of the evil flavour of a ' pheasant ' that one of his chums had shot near a native village, and had, much to the astonishment of the servants, brought home to be cooked and partaken of as a game-bird." There is an allied species of crow-pheasant, which is still more like a long-tailed game-bird, so much so that it is known to zoologists as *Centropus phasianus*. Here, then, we have examples of cuckoos which resemble other species and suffer in consequence. What have those naturalists who declare that mimicry is due to natural selection to say to this ?

The crow-pheasant is an easy bird to identify. The wings are chestnut in colour, while all the remainder of the plumage is black with a green or purple gloss.

But for the fact that the brown wings do not match well with the rest of the plumage, I should call the coucal a handsome bird. This, however, is not " Eha's " view.

The crow-pheasant is widely distributed in India, being found in gardens, in cultivated fields, and in the jungle. All the bird demands is a thicket or hedgerow in which it can take cover when disturbed. It does not wander far from shelter, for it is a poor flier. Its diet is made up chiefly of insects, but not infrequently it captures larger quarry in the shape of scorpions, lizards, small snakes, and the like delicacies. Probably fresh-water mollusca and crustacea do not come amiss to the bird, for on occasions I have seen it wading in a nearly dried-up pond. It certainly picks much of its food from off the ground, but, as it is often seen in trees, and is able to hop from branch to branch with considerable address, I am inclined to think that it sometimes feeds on the caterpillars and other creeping things that lurk on the under surface of leaves. I have never actually observed it pick anything off a leaf, for the coucal is of a retiring disposition. Like some public personages, it declines to be interviewed.

Its call is a very distinctive, sonorous *Whoot, whoot, whoot,* and, as the bird habitually calls a little before dawn in the early part of the hot weather, its voice is doubtless often attributed to some species of owl.

The nest is, we are told, globular in shape, considerably larger than a football, composed of twigs and grass and lined with dried leaves. The entrance consists of an aperture at one side. I must confess that I have not yet seen any of the creature's nests. I have located several, but each one of these has been placed in the midst of a dense thicket, which, in its turn, has been situated in the compound of one of my neighbours. The only way of bringing a nest built in such a position to human view is by pulling down the greater part of the thicket. This operation is not feasible when the thicket in question happens to be in the garden of a neighbour.

Large though the nest is, it is not sufficiently commodious to admit the whole of the bird, so that the long tail of the sitting crow-pheasant projects outside the nest. " When in this position," writes Hume, " the bird is about as defenceless as the traditional ostrich which hid its head in the sand." This remark would certainly be justified were the crow-pheasant to build its nest in mid-desert, but I fail to see how it applies when the nest is in the middle of a thicket into which no crow or other creature with tail-pulling propensities is likely to penetrate. " In Australia," continues Hume, " the coucal manages these things far better. There, we are told, ' The nest, which is placed in the midst of a tuft of grass, is of a large size, composed of dried grasses, and is of a domed form, with two openings, through one of which the head of the female protrudes while sitting, and her tail through the other.' On the other hand, the Southern Chinese coucal, which Swinhoe declares to be identical with ours, goes a step further, and gets rid of the dome altogether."

Young crow-pheasants are of exceptional interest. Three distinct varieties have been described. In some the plumage is barred throughout. Jerdon supposed that these are all young females. Other young birds are like dull-coloured adults ; these are smaller than the barred forms, and sometimes progress by a series of hops, instead of adopting the strut so characteristic of the species. These dull-coloured birds are very wild, whereas the barred ones are usually easily tamed. This interesting fact was pointed out by Mr. Frank Finn in his delightful volume *Ornithological and Other Oddities*. Jerdon regards these as young cocks. The third variety is coloured exactly like the adult. Finn does not accept Jerdon's view, for, as he points out, the three forms differ in habits, and the barred and dull-coloured forms do not appear to occur in the same brood ; the young in any given nest are either all barred, or all dull-coloured, or all like the adults in colour. So that if the barred and dull-plumaged birds represent different sexes, then all the individuals of a brood must be of the same sex. Instances of this phenomenon have been recorded, but they appear to be very rare. Finn therefore thinks that the three varieties of young correspond to three races. In this connection it is of interest to note that Hume divided this species into three : *Centropus rufipennis,* found in the Indian Peninsula and Ceylon ; *C. intermedius,* which occurs in Eastern Bengal, Assam, and Burma ; and *C. maximus,* that inhabits Northern India and Sind. Blanford, while uniting all these into one species, says, " unquestionably these are all well-marked races."

Finn had brought to him in Calcutta barred and dull-coloured young birds, these possibly correspond to the rufipennis and intermedius races. The matter needs further investigation.

In this connection it should be noted that the young of the Indian koel (*Eudynamis honorata*)—a cuckoo parasitic on crows—are of three kinds. Some are all black like the cock, some are barred black and white like the hen, while others, though nearly altogether black, display a few white bars. The fact that I have seen specimens of all three kinds of koel nestling in one garden at Lahore would seem to militate somewhat against the theory that these correspond to different races or *gentes.*

I have discoursed at such length on the crow-pheasant that our other respectable cuckoos will not receive adequate treatment. The interesting malkohas will not get an innings to-day. I trust they will accept my apologies.

I must content myself in conclusion with a few words regarding the *sirkeer* or grey ground-cuckoo. The scientific name of this species—*Taccocua leschenaulti*—affords an excellent example of the heights to which our scientific men can rise in their sublimer moments. This cuckoo always appears to me like a large babbler.

It has the untidy appearance, the sombre plumage, and the laborious flight of the " seven sisters." But it does not go about in flocks. It appears to consider that " two is company, three is none." Its cherry-red bill is the one bit of bright colour it displays. From its beak it derives its vernacular name *jungli tota* (jungle parrot), the villagers being evidently of opinion that the beak makes the parrot. This cuckoo seems to feed entirely on the ground, picking up insects of all sorts and conditions. It is found only in the vicinity of trees. In the Basti district of the United Provinces, where it is unusually abundant, I noticed it at almost every camping-ground I visited. Mango topes appear to be its favourite feeding-places. When alarmed it used to fly to the nearest cornfield, where it was quickly lost to view. Its habits are in many ways like those of the coucal. It builds a rough-and-ready nest, a mere collection of twigs with a few leaves spread over the surface. The eggs are chalky white, like those of the crow-pheasant. Both the cock and the hen take part in incubation.

It is a bird concerning the habits of which there is comparatively little on record. It therefore offers a fine field for the investigations of Indian ornithologists.

III

THE BROWN ROCK-CHAT

THE standard books on Indian ornithology give inaccurate accounts of the distribution of some species of birds. In certain cases the mistakes are due to imperfect knowledge, in others it is probable that the range of the species in question has undergone change since the text-books were published. There must of necessity be a tendency for a flourishing species to extend its boundaries. Growing species, like successful nations, expand. A correspondent informs me that the Brahminy myna (*Temenuchus pagodarum*) is now a regular visitor at Abbottabad and Taran Taran in the Punjab, whereas Jerdon states that the bird is not found to the west of the United Provinces. Similarly, there is evidence that the red turtle dove (*Œnopopelia tranquebarica*) is extending its range westwards. Oates states that the tailor-bird (*Orthotomus sutorius*) does not occur at elevations over 4000 feet, but I frequently saw it at Coonoor, 2000 feet higher than the limit set by Oates.

The brown rock-chat (*Cercomela fusca*) is another species regarding the distribution of which the text-books are in error. Jerdon gives its range as " Saugor,

Bhopal, Bundlekhand, extending towards Gwalior and the United Provinces." Oates says, "The western limits of this species appear to be a line drawn from Cutch through Jodhpur to Hardwar. Thence it extends to Chunar, near Benares, on the east, and to Jubbulpur on the south, and I have not been able to trace its distribution more accurately than this." Nevertheless, this bird is very abundant at Lahore, some two hundred miles north-west of the occidental limit laid down by Oates. Brown rock-chats are so common at Lahore, and the locality seems so well suited to their mode of life, that I cannot think that the species is a recent addition to the fauna of the Lahore district. It must have been overlooked. It is scarcely possible for one individual to have a personal knowledge of all parts of so extensive a country as India : we cannot, therefore, expect accuracy in describing the range of birds until an ornithologist does for each locality what Jesse has done for Lucknow, that is to say, compiles a list of birds observed in a particular neighbourhood during a period of observation extending over a number of years.

Let us now pass on to the subject of this essay. The brown rock-chat is a dull-reddish-brown bird, slightly larger than a sparrow. There is no outward difference between the cock and the hen, both being attired with quaker-like plainness. They are, however, sprightly as to their habits, being quite robin-like in behaviour. As they hop about looking for food they make every now and again a neat bow, and by this it is easy to identify them. They seem invariably to

c

inhabit dry, stony ground. Round about Lahore numbers of ruined mosques and tombs exist, and each of these is the home of at least one pair of brown rock-chats. But these birds by no means confine themselves to old ruins. They are very partial to plots of building land on which bricks are stacked. When a man determines to build a bungalow in Lahore he acquires a plot of land, and then has pitched on to it a quantity of bricks in irregular heaps, each heap being a cartload. These bricks are then left undisturbed for any period up to ten years. Among these untidy and unsightly collections of building material numbers of brown rock-chats take up their abode. But there are not enough ruins and collections of bricks to accommodate all the rock-chats of the locality ; consequently, many of them haunt inhabited buildings, and display but little fear of the human possessors of these. Indeed, an allied species (*Cercomela melanura*) is thought by some to be the sparrow of the Scriptures.

A cock rock-chat used at the beginning of each hot weather to come into the skylight of my office at Lahore and sing most sweetly, while his mate was sitting on her eggs hard by. As I had not then seen a nest of this species I sent a Mohammedan *chaprassi* into the Shah Chirag—a tomb in the office compound—to ascertain whether the nest was inside it or not. He brought back word that the nest was inside the sepulchre, but that Christians were not allowed inside, adding, however, that the fakir in charge thought that an exception might be made in my favour. A rupee settled the question. Matting was laid down so that

the saint's burying-place might not be defiled by the
dust that fell from the boots of the infidel, and a ladder
was taken inside. Let into the walls of the tomb were
a number of large niches. In one of these, of which the
base was some ten feet above the level of the ground,
was the nest of the brown rock-chats, containing three
beautiful pale blue eggs, blotched with light yellow
at the broad end. The ledge on which the nest was
built was covered with dust and pieces of fallen plaster,
which had evidently been accumulating there for
generations. The fallen plaster served as a foundation
for the little nursery, which was composed entirely of
fine dried grass. This had the appearance of being
woven into a shallow cup, but I am inclined to think
that the material had been merely piled on to the ledge,
and that the cavity had resulted from the sitting of the
bird. The nest was bounded on two sides by the wall,
and the part of it next to the wall was deeper than the
remainder. There was no attempt at weaving or
cementing, and the whole was so loosely put together
that it could have been removed only by inserting a
piece of cardboard under it, and thus lifting it bodily
away. In other niches were three disused nests,
one of which I appropriated ; they had probably been
made in previous years by the same pair of birds.
I subsequently came across another nest inside an
inhabited bungalow at Lahore, and another on the
inner ledge of the window of an outhouse. Hume
stated that a pair of brown rock-chats built regularly
for years in his house at Etawah. They do not invariably construct the nest inside buildings. Hume

writes: " Deep ravines and earthy cliffs also attract them, and thousands of pairs build yearly in that vast network of ravines that fringes the courses of the Jumna and Chambul from opposite Agra to Calpee. Others nest in quarries, and I got several nests from those in the neighbourhood of Futtehpoor Sikri."

During the nesting season the brown rock-chat knows not what fear is. Mr. R. M. Adam gives an account of a pair which built a nest in a hole in a bath-room wall. The birds did not appear to be frightened by people entering and leaving the room. When the first brood had been reared the hen laid a second clutch of eggs, and, on these being taken, she immediately laid a third batch. Colonel Butler writes: "During the period of incubation both birds are extremely pugnacious, and vigorously attack any small birds, squirrels, rats, lizards, etc., that venture to approach the nest." The tameness of the brown rock-chat, together with his alluring ways and sweet song, make him an exceptionally fascinating little bird.

IV

THE SCAVENGER-IN-WAITING

THE number of kites to be seen in any given place depends almost entirely upon the state of sanitation in that place. In England conservancy arrangements are so good that the kite is practically extinct. We have no use for the bird at home. "*Il faut vive,*" says the kite, "and if you do not provide me with offal I shall prey upon poultry," "As to your living," replies the farmer, "*Je n'en vois pas la necessité,* and, if you attack my poultry, I shall attack you." The kites in the United Kingdom were as good as their word ; so were the farmers. The result is that the kite is a *rara avis* at home ; a nestling born in the British Isles is said to be worth £25.

India teems with kites (*Milvus govinda*) ; we may therefore infer that sanitation out there is primitive. Unfortunately, we Anglo-Indians do not require the kites to enable us to appreciate this fact. Kites, however, are useful in giving us the measure of the insanitariness of a town. Lahore is a great place for kites. That city contains a greater proportionate number of these scavenger birds for its size than any other city or town I have ever visited. They are nearly as abundant as the crows ; further, that beautiful bird, commonly known as Pharaoh's chicken (*Neophron percnopterus*), shows his smiling face to one at every turn. Let me here observe that I am not calling anyone names ; I am merely stating a fact. If the Lahore municipal authorities take my words to heart, so much the better !

Kites are the assistant sweepers to Government ; I was going to say "honorary sweepers," but that would not have been strictly accurate, for in India nothing is done for nothing. The kites receive no money wages, nothing that comes under the Accountant-General's audit, but they are paid in truck. They are allowed to keep the refuse they clear away. This seems on the face of it to be a *bandobast* most favourable to the Government, a very cheap way of securing servants ; but, like many another arrangement which reads well on paper, it is in practice not so advantageous as it appears. Thus the kite is apt to put a wide, I might almost say an elastic interpretation on the word "refuse." To take a concrete example : the other day one of these birds swooped down and carried off the chop that was to have formed the *pièce de résistance* of my breakfast.

But, notwithstanding his many misdeeds, the kite is a bird with which we in India could ill afford to dispense, for he subsists chiefly upon garbage. Fortified with this knowledge, we are able to properly appreciate the sublime lines of the poet Hurdis :

> " Mark but the soaring kite, and she will reade
> Brave rules for diet ; teach thee how to feede ;
> She flies aloft ; she spreads her ayrie plumes
> Above the earth, above the nauseous fumes
> Of dang'rous earth ; she makes herself a stranger
> T' inferior things, and checks at every danger."

Now, I like these lines. Not that I altogether approve of the sentiments therein expressed. I would not advise anyone, not even a German, to learn table manners from the kite. What I do like about the above is the splendid manner in which the poet strikes out a new line. [N.B.—The poets and their friends are strongly advised to omit the forty lines that follow.] The vulgar herd of poets can best be compared to a flock of sheep. One of them makes some wild statement about a bird, and all the rest plagiarise it. Not so Hurdis ; he is no slavish imitator. He obviously knows nothing about the kite, but that is a trifle. If poets wrote only of things with which they were *au fait*, where would all our poetry be ?

What Hurdis did know was that, as a general rule, when you want to write about a bird of which you know nothing, you are pretty safe in reading what the poets say about it, and then saying the very opposite. That in this particular case the rule does not hold good is Hurdis's misfortune, not his fault. The kite happens to be almost the only bird about which the poets write correctly. This is a phenomenon I am totally unable to explain.

Cowper sang :

> " Kites that swim sublime
> In still repeated circles, screaming loud."

Writes Clare :

> " Of chick and duck and gosling gone astray,
> All falling preys to the sweeping kite."

King says :

> " The kite will to her carrion fly."

The most captious critic could not take exception to any of these sentiments. He might certainly pull a long face at Macaulay's

> " The kites know well the long stern swell
> That bids the Roman close."

But he would find it exceedingly difficult to prove that the kites do not know this.

But let us leave the poets and return to the bird as it is, for common though he be in the East, the " sailing glead " is a bird that will repay a little study. His powers of flight, his ability to soar high above the earth, to sail through the thin air with outstretched and apparently motionless wings, are equalled by few birds. Watch him as he glides overhead in great circles until he disappears from sight. He constantly utters his tremulous, querulous scream—*Chēē-hēē-hēē-hēē-hēē ;* his head is bent so that his beak points downwards, and few things are there which escape his keen eye. Suddenly he espies a rabble of crows squabbling over a piece of meat. Quick as thought he is full on his downward career. A second or two later the fighting, squawking crows hear the swish of his wings—a sound very familiar to them—and promptly make way for him. None desires to feel the grip of his powerful talons. He sweeps above the bone of contention, drops a little, seizes it with his claws, and sails away to the nearest housetop, where he devours his booty, fixing it with his talons as he tears it with his beak.

Crows love not the kite. His manner of living resembles theirs so closely that a certain amount of opposition is inevitable. Then, again, the kite never makes any bones about carrying off a young crow if the opportunity presents itself. If the truth be told, the crows are afraid of the kite. They will, of course, not admit this. You will never get a crow to admit anything that may be used as evidence against him.

The crows regard kites with much the same feelings that the smaller boys at school regard the big, bullying boys. Those who know the ways of the *corvi* (and who is there in India that does not ?) will not be surprised to hear that they never lose an opportunity of scoring off a kite. There is no commoner sight than that of a brace of them, as likely as not aided and abetted by a king crow, chasing the fleeing glead, and endeavouring to pull a beakful of feathers out of his rump.

But crows prefer to worry the kite upon *terra firma*, for the latter is a clumsy bird when on the ground. He is so heavy that he can only waddle along, and, notwithstanding his great pinions, he experiences difficulty in raising himself off a level plain. Hence it is when a kite is resting, half asleep, upon the ground, that the " lurking villain crows " usually worry him. It requires at least two of the " trebledated birds " to do this with success. One alights in front of the victim and the other behind him. This apparently harmless manœuvre is quite sufficient to excite the suspicions of the kite. He turns his eyes uneasily from crow to crow, and, although he utters no sound, he is probably cursing his luck that he has not a visual organ at the back of his head. If he is a sensible bird he will at once fly off, in hopes that the perditious crows will not follow him. If he remains, the posteriorly situated crow takes a peck at his tail. He, of course, turns upon the aggressor, and thus gives the front bird the opportunity for which it has been waiting. Sooner or later the kite has to move on.

Kites are very fond of settling on the tops of posts, and on other spiky places ; this feature they share with crows, green parrots, blue jays, and other birds. I cannot bring myself to believe that such perches are comfortable ; but, just as a small boy will prefer balancing himself upon a narrow railing to sitting on a proper seat, so do birds seem to enjoy perching on all sorts of impossible places. Birds are like small boys in many respects. A kite, of course, enjoys one great advantage when he elects to rest upon such a perch : it is then impossible for "ribald" crows to come and squat to right and to left of him.

Kites are not migratory birds in most parts of India. It is said, however, that the kites leave Calcutta during the rains. I have never visited the " Queen of Indian cities," so I cannot say whether or not the kites act thus. Jerdon, Blanford, and Cunningham all declare that they do ; but Finn writes : "How such an idea could have arisen I do not know. I have always noticed kites in the rains, and have never heard that they were ever in the habit of leaving Calcutta then." The truth of the matter seems to be that when it rains very heavily the streets of the city on the Hooghly are washed comparatively clean, and all the food of the " sailing glead " is swept out into the country, so the kites go after it, but they return as soon as the rain stops.

The nesting season for the kites is at any time when they feel disposed to undertake the cares of the family. The books tell us that it begins in January. This is correct. Where they go wrong is in asserting that it ends in April. I should rather say that it ends in December. It is true, however, that in Northern India the greater number of nests are constructed in the first three months of the calendar year.

The completed nest is about the size of a football, and is an untidy mass of twigs, rags, mud, brickbats, and such-like things. It is usually placed high up in a tall tree, not quite at the top, on a forked branch. It is not a great architectural triumph, but it serves its purpose. Two eggs are usually laid. These have a white ground blotched with red or brown. Kites object to having their nest pried into, so that he who attempts to steal the eggs must not be surprised if the owners attack him.

V

INDIAN WAGTAILS

"What art thou made of?—air or light or dew?
—I have no time to tell you if I knew.
My tail—ask that—perhaps may solve the matter;
I've missed three flies already by this chatter."

I QUITE agree with Mr. Warde Fowler that wagtails are everything that birds should be. They
are just the right size; their shape and form are
perfect; they dress most tastefully; they display
that sprightliness that one looks for in birds; their
movements are elegant and engaging; their undulating
flight is blithe and gay; their song is sweet and cheery;
they are friendly, and sociable, fond of men and
animals, "not too shy, not too bold." They are, in
short, ideal birds.

I know of nothing more enjoyable than to sit
watching a wagtail feeding at the water's edge.

"She runs along the shore so quickly," writes a long-
forgotten author, "that the eye is hardly able to
follow her steps, and yet, with a flying glance, she
examines every crevice, every stalk that conceals her
reposing or creeping prey. Now she steps upon a
smoothly washed stone; she bathes and drinks—and
how becomingly, and with what an air! The very

nicest *soubrette* could not raise her dress more coquet-
tishly, the best-taught dancer not move with more
graceful *pas* than the pretty bather as she lifts her
train and dainty feet. Suddenly she throws herself,
with a jump and a bound, into the air, to catch the
circling gnat ; and now should be seen the beating of
wings, the darting hither and thither, the balancing
and the shakes and the *allegretto* that her tail keeps
time to. Nothing can surpass it in lightness. In fine, of
all the little feathered people, none, except the swallow,
is more graceful, fuller of movement, more adroit or
insinuating, than the wagtail."

Wagtails are essentially birds of the temperate zone.
They remind us of a fact that we who dwell in the
tropics are apt to forget, namely, that there are some
beautiful birds found outside the torrid zone.

Fourteen species of wagtail occur in India, but the
majority of them leave us to breed. They bring up
their families in cool Kashmir, on the chilly, wind-
swept heights of Thibet, or even in glacial Siberia, and
visit India only in the winter when their native land
becomes too frigid even for them.

Many of the migratory wagtails do not show themselves in the southern portion of the peninsula, being
rightly of opinion that the climate of Upper India is
not far from perfect during the winter months.

There is, however, one species—the most lovable
of them all—the pied-wagtail (*Motacilla maderas-
patensis*)—which has discovered that it is possible to
live in the plains of India throughout the year ; and,
having made this discovery, it has decided that the

troubles and trials of the hot weather are lesser evils
than the inconveniences and perils of the long migratory journey. The head of this species is black,
relieved by a white streak running through the eye;
the wings and tail are mostly black, and a bib—or is
" front " a more correct word ?—of similar hue is
usually worn. The under parts of the bird are white.

The pied-wagtail is common all over India. It is
particularly abundant in the city of Madras, where it
is to be seen everywhere—on the house-top, in the
courtyard, in shady garden, in open field, and on the
river bank in company with the soldiers who solemnly
fish in the waters round about the fort.

When in Madras I used to see almost daily one of
these birds perched on the telegraph wire that runs
across the Cooum parallel with the Mount Road bridge.
The bird seemed to spend most of the day in pouring
forth its sweet song. When sitting on the wire its tail
used to hang down in a most unwagtail-like manner,
so that the bird looked rather like a pipit. Pied-wag-
tails sometimes appropriate suitable parts of the
bungalow for nesting sites; when this happens the
human occupant has plenty of opportunity of studying
their ways.

The remaining thirteen species of wagtails are
merely winter visitors to the plains of India. Two or
three of these are to be seen feeding, during the cold
weather, on every grass-covered field, and at the edge
of every *jhil*. In the latter place wagtails are nothing
short of a nuisance to the man who is out after snipe,
for they have the habit of rising along with the snipe,

and the white outer-tail feathers invariably catch the eye. Many a snipe owes its life to the wagtail.

The four commonest of the migratory wagtails are, I think, the white (*Motacilla alba*), the masked (*M. impersonata*), the grey (*M. melanope*), and the grey-headed wagtail (*M. borealis*). The two latter are characterised by much bright yellow in the lower plumage, which the first two lack ; but I am not going to attempt to achieve the impossible by trying to describe the various species of wagtail. Owing to the fact that these birds, like ladies of fashion, are continually changing their gowns, it is very difficult to state the species to which an individual belongs without examining that individual feather by feather. You may see a dozen wagtails of the same species catching insects on your lawn, each of which differs markedly from all his companions. Most of us are satisfied with the knowledge that a given bird is a wagtail, and are able to enjoy the poetry and grace of its motion without troubling our heads about its scientific name.

VI

THE TEESA

*B*UTASTUR TEESA used to be called the white-eyed buzzard, but one day a worthy ornithologist discovered that the bird was not the genuine article, that its legs and its eggs betrayed the fact that it is not a true buzzard. Therefore a new name had to be found for the bird. In their search for this, naturalists have not met with great success. Indeed, the last state of the bird is worse than the first, for it is now known as the white-eyed buzzard-eagle. To the adjectival part of the name no one can take exception, because the white eye and a whitish patch of feathers on the back of the head are the most remarkable features of a rather ordinary-looking fowl. The name " buzzard-eagle," however, is most misleading. Although, as I have previously had occasion to state, eagles are not quite the noble creatures the poets have made them out to be, to suggest that *Butastur teesa* is one of them is to insult the whole aquiline community. Eagles, notwithstanding the fact that they sometimes eat carrion, attack, each according to the size of its talons, quarry of considerable size, and are, in consequence, the

32

terror of other birds. As Phil Robinson says of them,
" they stand in the sky as the symbol of calamity.
When they stoop to the earth it is a vision of sudden
death." To speak thus of *Butastur teesa* would be,
as Euclid says, absurd. The white-eyed buzzard is
almost contemptible as a bird of prey ; he is a raptorial
degenerate, a mere loafer.

In India one often sees a white-eyed buzzard, some
mynas, a pair of doves, several bee-eaters, one or two
king crows, and a roller, sitting, all in a row, on a telegraph wire within a few yards of one another ; the
first and the last, as likely as not, on the tops of the
telegraph poles, looking like pillar saints. Contrast
this state of affairs with what happens when a hawk
or a falcon appears on the scene. " Take to woodland,"
writes Phil Robinson, " and fill it with your birds of
beauty and of song; put your 'blackbird pipers in every
tree,' and have linnets ' starting all along the bushes.'
Let melody burthen every bough and every cloud
hold a lark. Have your doves in the pines, and your
thrushes in the hawthorn ; spangle your thistle-beds
with restless goldfinches, and your furze with yellow-
hammers. The sun is shining brightly, and the countryside seems fairly overflowing with gladness. But with
a single touch you can alter the whole scene ; for let
one hawk come skimming round that copse yonder,
and the whole woodland is mute in the moment. Here
and there shrill warning cries of alarm, and here and
there a bird dipping into the central covert of the
brake. But for the rest there might not be one winged
thing alive in all the landscape. The hawk throws a

D

shadow of desolation as it goes, its wings scatter fears on either side ; silence precedes it and gloom pursues."

Small birds fear the hawk and despise the *teesa*, because they know that the former is as swift and energetic as the latter is slow and lazy. But it is not easy to understand why the white-eyed buzzard does not prey upon wild birds, because its wings are, in proportion to its size, longer than those of most birds of prey. It is not that *Butastur* considers birds unfit to eat. On the contrary, says Mr. C. H. Donald, " that he would love to catch a bird for his dinner is proved by the fact of his coming down to a bird behind a net as soon as he sees it, but I suppose experience has taught him that it is no use his trying to catch one in its wild state, and in full possession of its wings and feathers, and, consequently, he never tries." Thus, we have no alternative but to regard the white-eyed buzzard as a degenerate, a bird that might starve in the midst of plenty.

When a hungry *Butastur* sees flitting all around him potential meals in the shape of small birds, his feelings must be akin to those of the impecunious man in the comic song who, as he contemplates the insurance policy on the life of his shrewish wife, cries out : " Stone broke with fifty quid staring me in the face ! " The white-eyed buzzard has perforce to feed upon very humble quarry, upon the creeping and crawling things, upon beetles and insects, with an occasional rat or frog. His usual method of capturing his prey is very similar to that of the shrike, or butcher-bird, or, to come nearer home, to that of the true buzzards. He

takes up a position on a bare branch of a tree, a telegraph pole, a fence, or other point of vantage, such as a
heap of *kankar*, and there waits patiently until some
small creature wanders by. On to this he quietly drops,
secures it in his feeble talons, and returns to the perch
to devour his quarry and thus bring to a close one of
nature's little tragedies, of which millions are being
daily enacted. After he has finished his dinner he
loves to sit awhile, as the nursery rhyme tells us we
should do, and quietly digest what he has eaten. I
once disturbed a *Butastur* that had just finished a
heavy meal in the shape of a frog, with the result that
the bird " brought up " the frog !

Sometimes the white-eyed buzzard beats over the
ground in search of its quarry, but this is not his usual
modus operandi. If you would see the white-eyed
buzzard, go into an open place and watch for a brown
bird a little larger than a crow, sitting motionless on
some point of vantage, like Patience on a monument.
By its sluggish habits, its small size, its white eye, and
the whitish patch at the back of its head, you may
recognise it. It utters a peculiar plaintive screaming
call, which is heard mostly at the nesting season.
" In February and March," writes Mr. Donald, " just
before the breeding season, these birds may be frequently seen soaring high up in the heavens, and giving
vent to their plaintive call, and might be taken for
falcons if it were not for their much more rounded
wings. When at a height their breasts appear dark
and their wings (lower surface) very light and silvery."

Needless to say, the nest of this species is not a very

skilfully constructed affair. It is not more beautiful than a *dak* bungalow, but, like the latter, serves the purpose for which it is built. It is very like that of the common crow—a loosely-put-together collection of sticks, devoid of anything in the form of lining, and placed fairly high up in the fork of a tree. The tree selected is usually one with rather dense foliage, and one of a clump or row, in preference to a solitary tree ; nevertheless, I have seen a nest in an isolated tree. The eggs, which are greyish white, are not laid until some time after the nest has been made ready. *Teesas* are very noisy at the nesting season ; the sitting hen utters constantly a mewing cry, which renders the nest easy to locate ; but her vocal efforts pale into insignificance before those of the young hopefuls. These, to quote Mr. Benjamin Aitken, " keep up an incessant screaming for days before and after they leave the nest ; so that you cannot pass within two hundred yards of a brood of nearly fledged or newly fledged birds without being made painfully aware of their existence and good spirits."

VII

FALCONRY IN INDIA

LEST the title of this chapter should lead the reader to indulge in expectations that will not be realised, let me hasten to say that, in my opinion, hawking is a much overrated pastime. This statement will, of course, rouse the ire of the keen falconer, who will tell me that hawking is the sport of kings, and that it has no equal. To such a defence of the sport the obvious reply is that it has almost entirely died out in England, and that in India, where there is every facility for it, very few Europeans care to indulge in it. In Persia and India falconry is carried on in precisely the same way as it used to be in England. There can be little doubt that the sport originated in the East, and was introduced into the British Isles in Anglo-Saxon times. The hoods, the jesses, the lures, the gauntlets that are used in India to-day are exactly like those portrayed in old English hawking prints.

Hawks fall into two classes, according to the method of catching their quarry. These may be compared respectively to sprinters and long-distance men among human athletes. They are known to falconers as the short-winged or yellow-eyed hawks and the long-winged or dark-eyed hawks. The former adopt what I may perhaps call slap-dash methods. A furious rush is made at the quarry, and if this be not secured at the first onslaught the chase is given up. The second class adopts the slow but sure method. The falcon, having sighted its quarry, settles down to a long pursuit, keeps on and on until it finds itself above its victim, on to which it stoops. The second class of raptorial birds, which includes all the falcons, affords the better kind of sport, because the following of the chase entails some hard riding. For falconry of this kind a stretch of flat, open country is a *sine qua non*, and, as this is comparatively easy to find in India, one would naturally expect that the long-winged form of falconry would be the most popular among Indians. But this is not so. In Northern India, at any rate, that species of falconry that does not involve hard riding on the part of the falconer is the most practised. The gos-hawk (*Astur palabarius*) is the hawk most commonly used.

Perhaps the best method of conveying an idea of falconry to one who has not witnessed the sport is to describe a day's actual experience. The month is December, and the place Oudh. This means a sunny but perfectly cool day, so that riding, even when the sun is at its zenith, is delightful. Our party consists of an Indian gentleman—a Sikh and a large land-holder—who owns the hawks, and three Europeans all well mounted, also the chief falconer, indifferently mounted, who carries on his gloved forearm a goshawk.

Then there are two other falconers on foot, one carrying
a goshawk and the other a sparrow-hawk (*Accipiter
nisus*). Half a dozen beaters and three mongrel terriers
complete the party. The sparrow-hawk is hooded,
while the goshawk is not, being of a less excitable
nature. The hood is a leather cap, constructed so as
to cover up the wearer's eyes but not her beak. The
hood terminates in a point like a helmet. In the
summit some plumes are stuck, so that the hooded
bird has a fantastic appearance. Sparrow-hawks and
peregrines are made to wear these hoods when taken
out, until the falconer espies quarry, when he unhoods
his hawk and lets the ends of the jesses go. The jesses
are short straps made of soft leather, which all trained
hawks and falcons always wear. The goshawks are
both females. In all species of the *raptores*—listen to
this, ye suffragettes!—the female is larger and bolder
than the male, and hence is more highly esteemed by
the falconer. The female goshawk is known as a *baz*,
and is worth anything up to Rs. 150, while the male,
called the *jurra*, will never fetch more than Rs. 80.
The goshawks whose exploits I am about to recount
cost Rs. 80 and Rs. 60 respectively. They have been
trained more especially to take peafowl.

The party sets out in a southerly direction across an
uneven plain, much intersected by dried-up water-
courses. There is no cultivation on the plain, which
is to a large extent covered with long *sarpat* grass and
other xerophilous plants. We move along in an
irregular line, the dogs and beaters doing their best
to put up game. Suddenly a quail rises. " Let loose

the sparrow-hawk," cries the Sirdar. But, alas, the man carrying that bird has lagged behind, so the quail escapes. I may here say that on nine occasions out of ten when out hawking the man with the proper hawk is not where he should be. We continue our course, and presently come to a narrow river running through a deep *nullah*. Here two or three cormorants come flying overhead. They are forthwith "spotted" by the goshawks, which have all the time been eagerly looking about them in all directions. Having seen the cormorants, they begin tugging excitedly at their jesses. The falconers liberate the goshawks, and away they go in pursuit. After flying about eighty yards, first one goshawk, then the other, gives up the chase, and each repairs to the tree that happens to be nearest it. Then the falconers go up and show the birds pieces of meat, in order to entice them back to the fist. One *baz* immediately flies to the bait. Not so the other. She sits perched in her tree with an air of *j'y suis, j'y reste*. In a few seconds some crows catch sight of her and proceed to mob her by flying around her and squawking loudly. However, not one of them dares to touch her. Presently she too flies to her trainer, and the party moves on.

We next ford the river. On the far side the country is still more rugged, but contains more trees. Presently there is a great commotion in the thicket, and up gets a great peacock. The goshawks are again released and give chase. They fly low and make straight for the peacock, upon which they gain rapidly. We ride hard after them. After a flight of perhaps two hundred yards the hawks, when close up to the object of their chase, give up the pursuit, and fly to trees hard by. I ask their owner why they did not secure the peacock. He replies : " They would have taken it had it been a hen. They are not used to the male bird. Alas, my best hawk, which would take the cock, died last week ! " Let me here remark that I have never yet come across a falconer whose best hawk had not recently died. This is the inevitable excuse for the apparently invariable failure of the falcon to secure its quarry. To cut a long story short, neither of those goshawks secured anything that day. Later, the sparrow-hawk was sent after an unfortunate myna (*Acridotheres tristis*), which it secured after a chase of perhaps a dozen yards. Its talons struck the myna in the neck, and it soon killed it, not, however, before the poor little creature had emitted some heart-rending shrieks. The goshawk must occasionally catch something, or it would not fetch so large a price, and would not be so popular a bird with falconers in Northern India, but I imagine that on most days the hawking party returns without having secured anything.

Let me now give a brief account of hawking with the *Bhairi*, or peregrine falcon (*Falco peregrinus*). The scene, this time, is a huge expanse of flat plain in the Punjab, near the River Jhelum. The hawks belong to a European. We have ridden for several hours, not having succeeded in putting up quarry of any kind. As the falconer seems to have anticipated this, and as he has with him on trial a new peregrine, which he wants to see at work, an unfortunate crow, which was captured in the morning and has been carried round in a bag with us, is let go. He flies in a very stiff manner. When he has flown some eighty yards the peregrine is unhooded and let go after him. She at once flies upwards, and in a few seconds is above the crow, who, seeing her, drops to the ground and lies there on his back prepared to show fight. The falcon stoops at him, but seems to be afraid to tackle him on the ground. The falconer then runs up to him and tries to make him get up ; but he refuses, so he is picked up and thrown into the air. The falcon at once stoops at him, but before she reaches him the crow has again dropped to the ground, and still the falcon refuses to close with him. " That bird is of no use," is the comment of my host, an assertion which I do not feel inclined to contradict.

The only other kinds of falconry I have witnessed are those with hawk-eagles (*Spizætus nepalensis*), shikras (*Astur badius*), and merlins (*Æsalon regulus*). Hodgson's hawk-eagle is so large a bird that to watch him dashing after his quarry is a fine sight. It is said that this species can be trained to capture chinkara (*Gazella bennetti*). However, I have only seen it in pursuit of a hare that had been previously caught and then let loose. The hawk-eagle overtook this before it had gone fifty yards.

Hawking with the shikra is, in my opinion, very poor sport, for this little bird of prey makes but one dash at its quarry, and at once desists if this does not enable it to overhaul it. It is usually flown at quails or mynas. While waiting for its victim it is carried on the hand, but is not hooded. When one of the kind of bird to which it has been trained is flushed, the hawker takes the shikra in his hand, holds it between his thumb and fingers, and then throws it like a javelin in the direction of its quarry. Thus it enjoys the benefit of a flying start, but, notwithstanding this, it generally fails to make a catch.

The contest between a merlin and a hoopoe is an exceedingly pretty sight. The hoopoe is not a very rapid flier, but he is a past master in the art of jinking and dodging, and the manner in which he times the onslaught of the merlin, and jerks himself a couple of inches to right or to left, is a sight for the gods. The merlin, thus cheated of his victim, is carried on by sheer force of momentum some sixty yards before he can turn for another dash at the hoopoe. Meanwhile the latter is steadily flapping towards cover. The merlin is no more successful in his second dash, nor in his third or his fourth ; on each occasion the hoopoe escapes, apparently by the proverbial hair's-breadth. A single merlin is usually not clever enough to capture the wily hoopoe, but when two of them act in concert they usually succeed in doing so.

Such, then, is falconry as I have seen it. I concede that my experience has not been great, but I have witnessed enough to enable me to understand how it is that shooting has almost entirely displaced it as a pastime.

The training of hawks is, of course, most interesting, and must be a very fascinating pursuit to those engaged in it. When once the hawk or falcon has been trained, it appears to me that the best of the fun is over.

The going out in search of quarry seems only an excuse for spending a day in the open on horseback under very pleasant conditions.

VIII

HAWKS IN MINIATURE

EVEN as the earth is overrun by dacoits, robbers, and highwaymen in all places where the arm of the law is not far-reaching and hard-striking, so is the air infested with bandits. These feathered marauders fall into three classes, according to the magnitude of their quarry. There are, first, the eagles, falcons, and hawks, which attack creatures of considerable size. Then follow the shrikes or butcher-birds—pocket editions of the raptores—which prey upon the small fry among reptiles, mammals, and birds, also upon the larger insects. Lastly come the fly-catchers, which content themselves with microscopic booty, with trifles that the larger birds of prey do not deem worthy of notice. These last are able to swallow their victims bodily. Not so the shrikes and birds of prey, whose quarry has to be devoured piecemeal, to be captured, killed, then torn to pieces.

Similarity of calling not infrequently engenders similarity of appearance. Swifts and swallows afford a striking instance of this. Alike externally, they are widely separated morphologically. So is it with the shrikes and the raptores. The earlier naturalists were misled by this outward likeness, and, in consequence, classed the swifts with the swallows and the shrikes with the falcons.

Many are the points of resemblance between the greater and the lesser bandits of the air. The ferocity of their mien is apparent to the most casual observer. Michelet speaks of the eagle as having a " repulsively ferocious figure, armed with invincible talons, and a beak tipped with iron, which would kill at the first blow." Even more sinister is the aspect of the shrike. The broad black streak that runs from the bill to the nape of the neck serves to accentuate the fierce expression of the eye. The American naturalist Burroughs speaks of the shrike as a " bird with the mark of Cain upon him. . . . the assassin of the small birds, whom he often destroys in pure wantonness, or to sup upon their brains."

Much has been written about the cruelty of birds of prey. Their calling is indeed a barbarous one ; they undoubtedly inflict much pain ; but these are not reasons why they should be spoken of as villains of the deepest dye, as criminals worthy of the noose. The bird of prey kills his quarry because it is his nature to do so. He regards his victims as so many elusive loaves of bread, made for his consumption, to be obtained for the catching. The fly-catcher holds similar views regarding his quarry. We should bear in mind that the average insectivorous bird kills in the course of his life a vastly greater number of living things than does the eagle. The robin, for example, has been known to devour two and a half times its weight in earthworms in a single day. Were the daily tale of its victims placed end to end they would form a wriggling line fourteen feet in length. Yet writers abuse the fierce and vicious eagle, while they belaud the gentle and good robin. Thus Michelet writes with typical romantic fervour : " These birds of prey, with their small brains, offer a striking contrast to the numerous amiable and plainly intelligent species which we find among the smaller birds. The head of the former is only a beak ; that of the latter has a face. What comparison can be made between these giant brutes and the intelligent, all-human bird, the robin redbreast, which at this moment hovers about me, perches on my shoulder or my paper, examines my writing, warms himself at the fire, or curiously peers through the window to see if the spring-time will not soon return ? "

Writing of this description is possibly very magnificent, but it is not natural history. What is sauce for the goose is sauce for the gander. If it is wicked of the falcon to devour a duck, I fail to see that it is virtuous of the robin to gobble up a worm.

But to return to the shrike. His beak is very falcon-like. The short, arched, upper mandible, with its pointed, downwardly-directed tip and strong projecting tooth, is a weapon admittedly adapted to the tearing-up of raw flesh. The butcher-bird waits for his quarry much as the buzzard does, sitting immobile on the highest branch of a bush or low tree, whence he scans the surface of the earth. Something moving on the

ground arrests his attention. In an instant he has
swooped and seized a grasshopper. A second later
he is back on his perch, grasping his victim. I have
already stated that shrikes feed upon small mammals,
birds and reptiles, and large insects. These last
make up by far the greater portion of his menu. Often
have I watched the smaller species of Indian shrike
obtaining a meal, but never have I seen any
of these capture anything larger than an insect.
Mr. W. Jesse says of the Indian grey shrike (*Lanius
lahtora*)—the largest of our species: "It feeds on
crickets, locusts, lizards, and the like. It may occasionally seize a sickly or a young bird, but I have
never actually seen it do so." Other observers have
been more fortunate. Thus "Eha" says: "Sometimes
it sees a possible chance in a flock of little birds absorbed
in searching for grass seeds. Then it slips from its
watch-tower and, gliding softly down, pops into the
midst of them without warning, and strikes its talons
into the nearest." Similarly Benjamin Aitken writes:
"The rufous-backed shrike, though not so large as
the grey shrike, is a much bolder and fiercer bird. It
will come down at once to a cage of small birds exposed
at a window, and I once had an amadavat killed and
partly eaten through the wires by one of these shrikes
which I saw in the act with my own eyes. The next
day I caught the shrike in a large basket, which I had
set over the cage of amadavats.   "On another occasion I exposed a rat in a cage for the purpose of
attracting a hawk, and in a few minutes found a *Lanius
erythronotus* fiercely attacking the cage on all sides."

I am disposed to regard such cases as the exceptions which prove the rule that the food of, at any rate, the smaller species of shrike, consists mainly of insects. This would explain why so few shrikes' " larders " are discovered in India. Every popular book on natural history describes how the butcher-bird, having killed his victim, impales it upon a thorn, and leaves it there to grow tender preparatory to devouring it. I have not been lucky enough to come across one of these larders. Other naturalists have been more fortunate, and we may take it as an established fact that even the smaller Indian species of butcher-birds sometimes impale their victims on thorns. The existence of such larders is easily accounted for. When the little butcher captures a victim so large that it has to be torn to pieces before consumption, he has to find some method of fixing it while tearing it up. He is not heavy enough to pin it to the ground with his talons, as a raptorial bird does, so must perforce utilise the fork of a tree or a large thorn. Having taken his fill, he flies away, leaving the remains of his dinner impaled on the thorn, where it is discovered by some enterprising ornithologist.

Fifteen species of *Lanius* are described as existing in India. Of these the three most commonly seen are the rufous-backed, the bay-backed, and the grey species.

The rufous-backed shrike (*Lanius erythronotus*) is the only butcher-bird that is abundant on the Bombay side. It is about the size of a bulbul. It sits bolt upright, with tail pointing to the ground, and in this

E

attitude watches for its quarry. It has a grey head, with a conspicuous broad black band—the mark of the butcher-bird community—running through the eye. Its back is reddish brown. It has a white shirt-front, which makes it easy to see; moreover, it always sits on an exposed perch. To mistake a shrike is impossible. There is no other fowl like unto it.

The bay-backed species (*L. vittatus*) is a somewhat smaller bird, but is very like *erythronotus* in appearance. It may, however, be distinguished at a glance when on the wing by the white in the wings and tail.

The third common species—the Indian grey shrike (*L. lahtora*)—has the whole of the back grey, and thus is recognisable without difficulty.

The nest of the butcher-bird is an untidy, cup-shaped structure, from which pieces of rag frequently hang down. As often as not it is built in a thorny tree, and, by preference, pressed up close against the trunk. Baby shrikes make their *début* into the world during the hot weather.

IX

THE ROOSTING OF THE BEE-EATERS

ONE evening in August I was "on the prowl" with a pair of field-glasses, when I came across a tree from which emanated the twittering of many green bee-eaters (*Merops viridis*). As the sun was about to set, it was evident that these alluring little birds were getting ready to go to sleep. Most birds seem to roost in company. They do so presumably for the sake of companionship, warmth, and, perhaps, protection. To my mind there is no sight more amusing than that of a number of little birds going to bed, so I turned aside to watch these emerald bee-eaters. The tree in question was an isolated one, growing at the side of a field. I do not know its name, but it was about twenty feet high, with fairly dense foliage, the leaves being in colouring and shape not unlike those of the rose. The bee-eaters in the tree were making a great noise ; all were twittering at the top of their musical little voices, and, as there were certainly more than forty of them, to say nothing of some other birds, the clamour may be imagined. From a little distance it sounded like the calling of many cicadas. The birds were evidently busy selecting perches on which to pass the night, and there was, as there seems always to be on such occasions, a certain amount of squabbling. I was going to say " fighting," but perhaps that would be too strong a word to use for this scramble for places. At times, indeed, the scramble would develop into a fight, and two birds emerge snapping at one another. Once outside they would desist from fighting and return to the tree. Occasionally a bee-eater would dart out of the tree, and make a sally after some flying insect, and, having caught it with a loud snap of its mandibles, return to the tree and disappear into the " leafy bower." While this was going on amid the foliage, fresh bee-eaters kept coming in from a distance, mostly in pairs. These all made direct for the tree, evidently knowing it well.

I crept up to within about six yards of the dormitory, so as to witness as much as possible of what was going on amongst the leaves.

Some of the birds looked as though they had settled down for the night, since they were quite quiet. Two, in particular, had taken up a position, side by side, close up against one another on a somewhat isolated bough. They sat there quite still except for an occasional turn of the head, which seemed to express surprise and annoyance at the clamour of their fellows. Several other individuals had settled down in the same manner, in rows of two or more, huddled as close as possible together, each row being on a separate branch.

I noticed one line of eight bee-eaters, squeezed up against one another, and very pretty did the eight little heads look. But these rows were subjected to constant disturbance, and were continually being broken up and re-formed. The disturbances came both from within and from without. One of a row, usually the outside one (outside berths are not appreciated by the bird-folk), would suddenly determine to better his position, which he would seek to do by hopping on to his neighbour's back, and trying to wedge himself in between him and the next bird. This would be resented by the aforesaid neighbour, who would try to shake off the intruder, and the struggle that ensued would, as often as not, result in the break-up of the whole row. Birds that had not already found suitable perches would join rows already in existence. This was a constant source of disturbance. Perhaps four bee-eaters would be sitting on a bough which their weight caused to hang horizontally, then a fifth bird would take it into his little head to alight at the extreme tip of the branch, and bear it down to such an extent that those already on it had to grip hard to maintain their equilibrium. It must be very disconcerting and annoying to a sleepy little bird when the angle of its perch is suddenly changed by fifteen or twenty degrees !

While I was watching all this some village boys caught sight of me, and, with the curiosity so characteristic of the Punjabi, came up to see what I was looking at. Shortly after their arrival one of them showed his country manners by clearing his throat with such violence as to frighten all the bee-eaters out of

the tree in which they were settling down for the night !
Some flew to a neighbouring tree, but the majority
circled in the air with loud twitterings. Within less
than three minutes, however, all were back again,
trying to find suitable perches. Before they had half
settled down a boy again disturbed them. This was
obviously done to annoy me, so I sent the urchins
about their business. All the bee-eaters were back
again almost immediately. By this time the sun had
disappeared below the horizon, a fact which the birds
seemed to appreciate, judging by the celerity with
which they settled down. It soon grew so dark that I
could scarcely distinguish the birds from the foliage
which they resemble so much in hue. But for the
black streak through the eye I should not have been
able to do so. I now crept up under the tree, and was
able, by looking up, to distinguish little groups of bee-
eaters huddled together. I noticed several couples,
two rows of three, four rows of four, and one of five.
The tails projected from behind, and by counting these
I was able to determine the number in a row. I noticed
that the tails were not parallel ; some were crossed
by others, showing that the birds do not roost so closely
packed as they appear to be when looked at from the
front. Birds are composed largely of feathers, so that
it is easy for them to have the appearance of being
packed like sardines in a tin when in reality they have
plenty of room.

All the birds in a row faced the same way, but some
rows looked one way and others another.

Bee-eaters do not sleep with the head under the wing,

as some birds do, but are content to allow it to drop into their downy shoulders.

The little company did not all roost at the same elevation, but none slept on the lowest branches, nor could I distinguish any on the highest boughs. I should say that all the birds roost in the middle zone of the tree. The branches selected were not necessarily those where the foliage was thickest, nor, so far as I could make out, where the sleeping birds would be best protected from dew and rain. As it rained very heavily in the night in question, some of those bee-eaters must have had a nocturnal shower-bath.

X

OWLS

IT is the misfortune of owls that they are universally unpopular. They are heartily detested by their fellow-birds, who never miss an opportunity of mobbing them. They are looked upon with superstitious dread by the more ignorant classes all the world over. Jews and Gentiles, Christians and heathens, alike hate them. Owls are thought to be " death birds," " foul precursors of the fiend," " birds whose breath brings sickness, and whose note is death," death's dreadful messengers, Satan's *chapprassis*, the devil's poultry. Poets join with the vulgar *plebs* in showering abusive epithets upon them. Owls are gibbering, moping, dull, ghastly, gloomy, fearful, cruel, fatal, dire, foul, baleful, boding, grim, sullen birds, birds of mean degree and evil omen. The naturalist is, however, above the vulgar and ill-founded prejudice against the " sailing pirates of the night." To him, owls are birds of peculiar fascination and surpassing interest. They are of peculiar fascination because he has learned comparatively little about their habits. We day folk have but a slender knowledge of the lives of the creatures of the night. To

56

most of us owls are *voces, et præterea nihil*—voices
which are the reverse of pleasant. Owls are of surpassing interest to the naturalist on account of their
perfect adaptation to a peculiar mode of life.

The owl is a bird of prey which seeks its quarry by
night, a "cat on wings," as Phil Robinson hath it.
A master of the craft of night-hunting must of necessity
possess exceptional eyesight. His sense of hearing too
must be extraordinarily acute, for in the stillness of
the night it is the ear rather than the eye that is relied
upon to detect the presence of that which is sought.
Another *sine qua non* of owl existence is the power of
silent progression. Were the flight of owls noisy, like
that of crows and other large birds, their victims would
hear them coming, and so be able to make good their
escape. He who hunts in the night has to take his
quarry by surprise. Everyone must have noticed
the great staring orbs of the owl. Like the wolf in the
story of Little Red Riding Hood, it has large eyes in
order the better to see its victim. The eye of the owl
is both large and rounded, and the pupil is big for the
size of the eye in order to admit as much moonlight
as possible. The visual organs of the owl are made
for night work, and so are unsuited to the hours of
sunlight. Ordinary daylight is probably as trying to
the owl as the glare of the noonday sun in the desert
is to human beings. But it is not correct to speak of
the owl as blind during the day. He can see quite well.
He behaves stupidly when evicted from his shady
haunts in the daytime because he is momentarily
blinded, just as we human beings are when we suddenly

plunge from the darkened bungalow into the midday
sun of an Indian June. I have seen owls of various
species either sitting on a perch or flying about quite
happily at midday.

The chief reason why most owls are so strictly nocturnal is because they are intensely unpopular among
the birds of the day. These give them a bad time whenever they venture forth. In this the crows take the
lead. Crows, like London cads, are intensely con-
servative. They hate the sight of any curious-looking
or strangely dressed person. Put on a Cawnpore tent
club helmet, and walk for a mile in the East End of
London, and you will learn the kind of treatment to
which owls are subjected by their fellow-birds when
they venture forth by day. Mr. Evans, writing of the
owl in his volume, *The Songs of Birds*, says : " There
is some sad secret, which we do not know, which no
bird has yet divulged to us, and which seems to have
made him an outcast from the society of birds of the
day. He is branded with perpetual infamy." I trust
that Mr. Evans will not take it ill if I state that there is
no secret in the matter. Diurnal birds are not aware
that the country is full of owls, so that when one of
these appears they regard it as an intruder, a new
addition to the local fauna, to extirpate which is
their bounden duty. When a cockatoo escapes from its
cage the local birds mob it quite as viciously as they do
the owl.

Another peculiarity of the owl lies in the position of
its eyes. These are forwardly directed. In most birds
the eyes are placed at the side of the head, so that owls

alone among the feathered folk can truly be said to possess faces. The position of a bird's eyes is not the result of chance or accident. A creature whose eyes are forwardly directed can see better ahead of him than he could were they placed at the sides of the head, but he cannot see what is going on behind his back. Animals whose eyes are at the side of the head have a much wider range of vision, for the areas covered by their visual organs do not overlap. Such creatures cannot see quite so well things in front of them, but can witness much of what is going on behind them. They are therefore better protected from a rear attack than they would be did their eyes face forwards. The result of this is that, if we divide birds and beasts into those which hunt and those which are hunted, we notice that in the latter the tendency is for the eyes to be placed at the sides of the head. They thereby enjoy a wider range of vision, while in the former the tendency is for the eyes to be so situated as to enable them best to espy their quarry. Compare the position of the eyes in the tiger and the ox, in the eagle and the sparrow. The tiger and eagle have little fear of being attacked, so have thrown caution to the winds and concentrated their energies to equipping themselves for attack. In owls the eyes are more forwardly directed than in the diurnal birds of prey, because they have to hunt their quarry under more difficult conditions. Even when its ears inform the owl that there is some creature near by, it requires the keenest eyesight to detect what this is. The position of a bird's eyes is determined by natural selection. With colour and

such-like trifles natural selection has but little to do.
It works on broad lines. It determines certain limits
within which variations are permissible ; it does not go
into details. So far as it is concerned, an organism may
vary considerably, provided the limits it defines are
not transgressed. This statement will not meet with
the approval of ultra-Darwinians, but I submit that it
is nevertheless in accordance with facts. If we try to
account for every trivial feature in every bird and
beast on the principle of natural selection, we soon find
ourselves lost in a maze of difficulties.

It is because the eyes of owls are forwardly-directed
that they are such easy birds to mob. They can see
only in one direction—a limitation which day-birds
have discovered. The result is that when the owls do
venture forth during the daytime, they come in for
rough handling. The position of the eyes in the owl
would lead us to infer that the bird has but few enemies
to fear, and, so far as I am aware, there is no creature
which preys on them, except, of course, the British
gamekeeper. Why, then, are owls not more numerous
than they are in those countries where there are no
gamekeepers to vex their souls ? The population of
owls must of course be limited by the abundance of
their quarry. But there is more than enough food
to satisfy the hunger of the existing owls. What, then,
keeps down their numbers ? Mr. F. C. Selous has
asked a similar question with regard to lions in Africa.
Even before the days of the express rifle lions were
comparatively scarce, while the various species of deer
roamed about the country in innumerable herds. The

answer must, I think, be found in the intensity of the struggle for existence. Nature balances things with such nicety that the beasts of prey have their work cut out to secure their food. The quarry is there in abundance ; the difficulty is to catch it. If this be so, it follows that the weaker, the less swift, the less skilled of the predaceous creatures must starve to death. In that case the lot of birds and beasts of prey is a less happy one than that of their victims. These latter are usually able to find food in abundance, and death comes suddenly and unexpectedly upon them when they are in the best of health. How much better is such an end than death due to starvation ?

In most birds the opening of the auditory organ is small ; in owls it is very large and is protected by a movable flap of skin, which probably aids the bird in focussing sounds. In many species of owl the two ear-openings are asymmetrical and differ in shape and size. This arrangement is probably conducive to the accurate location of sound. Want of space debars me from further dilating upon the wonderful ear of the owl.

In conclusion, mention must be made of the flexible wing feathers, and their soft, downy edges. Air rushing through these makes no sound. Hence the ear may not hear, but

> " The eye
> May trace those sailing pirates of the night,
> Stooping with dusky prows to cleave the gloom,
> Scattering a momentary wake behind,
> A palpable and broken brightness shed,
> As with white wings they part the darksome air."

XI

A BUNDLE OF INIQUITY

THE common squirrel of India is a fur-covered bundle of iniquity. He is a bigger rascal than either the crow or the sparrow. I am aware that these statements will not be believed by many residents of Northern India. I am sorry, but the truth must be told. Let those who will imagine *Sciurus palmarum* to be a pretty, fluffy little creature, as charming as he is abundant. I know better. I have sojourned in Madras. In Northern India the little striped squirrel is merely one of the many tribes that live on your frontier ; in South India he is a stranger who dwells within your gates. We who are condemned to residence in the plains of Northern India keep our bungalows shut up during the greater part of the year in order to protect ourselves from the heat, or the cold, or the dust, or whatever climatic ill happens to be in season. And when the weather does permit us to open our doors we have to guard them by means of *chiks* from the hordes of insects that are always ready to rush in upon us. Thus we keep the squirrel at arm's length. In Madras you lead a very different life. The gentle breeze is always welcome, you rarely, if ever, close the doors of your bungalow, for extremes of temperature are unknown. Nor are you obliged to protect every aperture by means of a *chik*. There is thus no barrier between the squirrel and yourself. The result is that the impudent little rodent behaves as though he believed that men build their bungalows chiefly for his benefit. Not content with living rent-free in your house during the nesting season, he expects you to furnish his quarters for him, and to provide him with food. As I have hinted elsewhere, Indian bungalows are constructed in such a manner as to lead one to infer that there is a secret compact between the builders and the fowls of the air. The rafters rarely fit properly into the walls, and the spaces left make ideal nesting sites for sparrows and squirrels. These last, although devoid of wings, are such adepts at climbing that there are few spots in any building to which they are unable to gain access.

In Madras punkahs are up all the year round, and, as usually they are pulled only at meal times, squirrels regard them as paths leading to their nests. Running up the hanging rope, walking, Blondin-like, along the leathern thongs that lead to the punkah, jumping from these on to the top of the punkah frame, climbing up the rope to a rafter, and marching along this to the nest, are feats which the little striped rodent performs without effort.

In default of a suitable cavity, the squirrel constructs, among the branches of a tree, a large globular nest, which has the appearance of a conglomeration of grass, straw, and rubbish, but it contains a cosily lined central cavity. Any available soft material is used to make the interior of the nest warm and comfortable. When squirrels are nesting it is not safe to leave any balls or skeins of wool lying about the bungalow. The fluffy little creatures sometimes display considerable ingenuity in adapting materials for use in nest construction. One rascal of my acquaintance destroyed a nearly new grey *topi*, finding the felt covering and the pith " the very thing " for nest-lining.

Books on natural history inform us that the food of this species of squirrel consists of seeds, fruits, and buds, with an occasional insect by way of condiment. This is the truth, but it is not the whole truth. The above list does not by any means exhaust the menu of *Sciurus palmarum*. My experience shows him to be nearly as omnivorous as the myna. Occasionally I fall asleep again after my *chota hazri* has been brought. In Madras I was sometimes punished for my laziness by the disappearance of the toast or the butter. Needless to state that theft had been perpetrated, and that the crows and the squirrels were the culprits.

On one occasion I feigned sleep in order to see what would happen. For a little all was still ; presently a squirrel quietly entered the room, took a look round, then climbed up a leg of the table and boldly pulled a piece of toast out of the rack which was within a couple of feet of my face. It was no easy matter for the little thief to climb down the leg of the table with his big load. A loud thud announced that the toast had fallen on to the floor. The squirrel scampered away in alarm, leaving his booty behind him. In a few seconds his head appeared at the doorway ; having

regarded me attentively with his bright little eye, and
satisfied himself that all was well, he advanced to the
toast and bore it off. But, alas, the way of transgressors
is hard ! A "lurking, villain crow," who had been
watching the theft from the verandah, pounced upon
the thief, and bore off his ill-gotten toast. The wrath
of the squirrel was a sight for the gods. His whole
frame quivered as he told that crow what he thought
of him.

Sciurus palmarum is very fond of bread and milk, and
will, in order to obtain this, perform deeds of great
daring. I once kept a grackle, or hill-myna. This
bird, when not at large, used to dwell in a wicker cage.
In a corner of this cage a saucer of bread-and-milk
was sometimes placed. The squirrels soon learned to
climb up the leg of the table on which the cage stood,
insert their little paws between the bars, and abstract
the bread-and-milk, piece by piece. In order to
frustrate them, I placed the saucer in the middle of
the cage. Their reply to this was to gnaw through a
bar, and boldly enter the cage. They grew so audacious
that they used to walk into the cage while I was
present in the room ; but, of course, the least movement on my part was the signal for them to dash away
into the verandah. On one occasion I was too quick
for a squirrel who was feeding inside the grackle's cage.
I succeeded in placing my hand in front of the gnawed-
through bar before he could escape. He dashed about
the cage like a thing demented, and so alarmed the
myna that I had to let him out. In half an hour he
was again inside the cage !

F

The little striped squirrel feeds largely on the ground.
As every Anglo-Indian knows, it squats on its hind legs
when eating, and nibbles at the food which it holds
in its fore-paws. In this attitude its appearance is
very rat-like, its tail not being much *en évidence*. It is
careful never to wander far away from trees, in which
it immediately takes refuge when alarmed. It does
not always wait for the seeds, etc., upon which it feeds,
to fall to the ground : it frequently devours these while
still attached to the parent plant. Being very light,
it can move about on slender boughs. It is able to
jump with ease from branch to branch, but in doing so
causes a great commotion in the tree ; its arboreal
movements seem very clumsy when compared with
those of birds of the same size.

Squirrels are sociably inclined creatures ; when not
engaged in rearing up their families they live in colonies
in some decayed tree. At sunrise they issue forth from
the cavity in which they have slept, and bask for a
time in the sun before separating to visit their several
feeding-grounds ; at sunset they all return to their
dormitory. Before retiring for the night they play
hide-and-seek on the old tree, chasing each other in
and out of the holes with which it is riddled.

Young squirrels are born blind and naked, and are
then ugly creatures. Their skin shows the three black
longitudinal stripes—the marks of Hanuman's fingers—
which give this creature its popular name. The hair
soon grows and transforms the squirrels.

A baby *Sciurus* makes a charming pet. The rapid
movements are a never-failing source of amusement. It

is feeding out of your hand when it takes alarm at apparently nothing, and, before you can realise what has happened, it has disappeared. After a search it is found under the sofa, on the mantelpiece, or out in the garden. I know of one who took refuge in its owner's skirts. She had to retire to her room and divest herself of sundry garments before she could recover it. Once, in trying to catch a baby squirrel that was about to leap off the table, I seized the end of its tail ; to my astonishment the squirrel went off, leaving the terminal inch of its caudal appendage in my hand, nor did the severance of its note of interrogation seem to cause it any pain. A squirrel's tail, like a lamp brush, is composed mainly of bristles.

XII

THE INTERPRETATION OF THE ACTIONS OF ANIMALS

THE proper interpretation of the actions of animals is one of the greatest of the difficulties which confront the naturalist. We all know how liable a man's actions are to be misinterpreted by his fellow-men, whose thoughts and feelings are similar to his. How much more must we be liable to put false constructions on the acts of those creatures whose thoughts are not our thoughts and whose feelings are not our feelings? The natural tendency is, of course, to assign human attributes to animals, to put anthropomorphic interpretations on their actions, to endow dumb creatures with mental concepts like those of man—in short, to credit them with reasoning powers similar to those enjoyed by human beings. That this is incorrect is the opinion of all who have made a study of the question, and yet even such seem unable completely to divest themselves of the tendency to regard animals as rather simple human folk. I do not wish to speak dogmatically upon this most difficult subject. Let it suffice that it is my belief that animals do not possess the mental powers

68

popularly ascribed to them. My object is not to argue, but to record some instances showing how liable we are to misinterpret animal actions.

Some time ago, while walking near the golf-links at Lahore, I noticed a rat-bird, or common babbler (*Argya caudata*, to give it its proper name), with a green caterpillar hanging from its beak. The succulent insect was, of course, intended for a young bird in a nest near by. Being in no hurry, I determined to find that nest. Under such circumstances, the easiest way is to sit down and wait for the parent bird to indicate the position of the nursery. The bird with the caterpillar had seen me, so, instead of flying with it to the nest, moved about from bush to bush uttering his or her note of anger (I do not pretend to be able to distinguish a cock from a hen rat-bird). In a few minutes the other parent appeared on the scene, also with something in its beak. Observing that all was not well, it too began to " beat about the bush," or rather from one bush to another. Meanwhile, both swore at the ungentlemanly intruder. However, I had no intention of moving on before I found that nest. After a little time the patience of the second bird became exhausted ; it flew to a small bush, into which it disappeared, to reappear almost immediately with an empty beak. I immediately advanced on that bush, of which the top was not three feet above the ground. In the bush I found a neatly constructed, cup-shaped nest, which contained five young rat-birds. I handled these, taking one ugly, naked fellow in my hand in full view of the parents, who were swearing like bargees. I was careful

to make certain that the mother and father could see
what I was doing, for I was anxious to find out how
far their laudable attempts at the concealment of the
nest from me were due to the exercise of intelligence.
Having replaced the baby bird in the nest, I returned
to the place where I had waited for the parents to
direct me to their nursery, and watched their future
actions. If they had been acting intelligently, they
would reason thus, " The great ogre has found our
nest and seen our little ones. If he wants them we
are powerless to prevent him taking them. The
game of keeping their whereabouts hidden from him
is up. There is nothing left for us to do but to continue
to feed our chicks in the ordinary way without further
attempt at concealment." If, however, they were
acting blindly, merely obeying the promptings of the
instinct which teaches them not to feed their young
ones in the presence of danger, they would be as
unwilling now to visit the nest as they were after they
first caught sight of me. They pursued the latter
course, thus demonstrating that this seemingly most
intelligent behaviour is prompted by instinct.

It is a well-known fact that some birds, such as the
partridge, whose young are able to run about when
first hatched, behave in a very clever manner in
presence of danger. The mother bird acts as though
her wing was broken, and flutters away from the intruder with what appears to be a great and painful
effort. By this means she draws the attention of the
enemy to herself ; meanwhile her chicks are able to
hide themselves in whatever cover happens to be con-

venient. If anything looks like an intelligent act this surely does. But in this case appearances are deceptive. It sometimes happens that a hen partridge acts in this manner before her eggs are hatched. Under such circumstances the pretence of a broken wing is not only useless, but positively harmful, since it probably directs the attention of the intruder to her white eggs. This feigning of injury would thus appear to be a purely instinctive act, a course of behaviour dictated by natural selection. Mr. Edmund Selous discusses the origin of this peculiar habit in that admirable book entitled *Bird Watching*, to which I would refer those who are interested in the matter. Instances such as these, of acts which are only apparently purposeful, could easily be multiplied. They should prevent our rushing to the conclusion that because a cat, or dog, or horse behaves in a sensible manner under certain conditions, it is exercising intelligence. Natural selection has brought instinct to such perfection that many instinctive actions are very difficult to distinguish from those which are intelligent.

XIII

AT THE SIGN OF THE FARASH

THE farash tree (*Tamarix articulata*), regarded from the point of view of a human
being, is everything that a tree should not
be. Its wood has little or no commercial
value, being of not much use even as fuel. Its needle-
like leaves afford no shade. It has a dusty, dried-up,
funereal appearance. During the day it absorbs a large
amount of the sun's heat, which it emits, with interest,
at night-time, so that if, on a hot-weather evening, you
happen to pass near a farash tree you cannot fail to
notice that the temperature of the air immediately
surrounding it is considerably higher than it is elsewhere. Each farash tree becomes, for the time being,
a natural heating stove. In appearance the farash is
not unlike a stunted casuarina tree. It is what botan-
ists call a xerophile ; it is addicted to dry, sandy soil,
and is found only in the more desert-like parts of Sind
and the Punjab. The one redeeming feature of the
farash tree is the shelter it affords to the fowls of the air.
Its wood is so soft and so liable to decay that the tree
seems to have been evolved chiefly for the benefit
of those birds which nest in holes. The interior of

every aged farash is as full of cavities as a honeycomb. A grove of farash trees is a veritable bird hotel; it might with truth be called *L'Hôtel des Oiseaux*. Like many of the hotels built for the accommodation of human beings, the Farash Hotel is almost deserted at some periods of the year and overcrowded at others. It has its " season." During the winter months many of its rooms remain untenanted. The more commodious ones, however, are occupied all the year round; some by spotted owlets (*Athene brama*), and others by the little striped squirrel (*Sciurus palmarum*). The spotted owlets do not, like most birds, visit the farash merely for nesting purposes; they live in it, lying up in their inner chamber during the day, immune from the attacks of crows, kites, drongos, and other birds that vex the souls of little owls. No matter at what season of the year you call at the hotel, you will find Mr. and Mrs. Spotted Owlet at home during the daytime. If you tap on the trunk, which is tantamount to knocking at the door or shouting " *Koi hai*," you may expect to see appear at the door of the suite occupied by the owlets a droll little face, that will bow to you, but with such grimaces as to leave no doubt that you are unwelcome.

The squirrels are winter residents in the hotel; they like to dwell in it throughout the year, but are not always permitted to do so. Numbers of them are ejected every February by the green parrot (*Palæornis torquatus*). The green parrot is a bully, and is neither troubled by the Nonconformist conscience, nor hampered by the Ten Commandments; so that, when he has set his heart on a certain suite in the hotel, he proceeds to install himself therein, regardless of the vested interests of the squirrels. The " season " may be said to begin with the arrival of the green parrots. These rowdy creatures make things " hum," and must cause considerable annoyance to the more respectable birds that stay in the hotel. The green parrot is to bird gentlefolk what the Italian organ-grinder is to the musical Londoner—an ill that has to be endured. The little coppersmith (*Xantholæma hæmatocephala*) takes up its quarters in the bird hotel early in the season. It is very particular as regards its accommodation. It holds, and rightly holds, that rooms which have already been lived in are apt to harbour parasites and carry disease, so insists on hewing out a chamber for itself. Owing to the industry of both the cock and the hen, the excavation of their retort-shaped nesting chamber occupies surprisingly little time, and the neat, circular front-door that leads to it compares very favourably with the irregular, broken-down-looking entrance to the quarters occupied by the parrots or owlets. As often as not the coppersmith excavates its nest in a horizontal bough, in which case the entrance is invariably made on the under surface, with the object of preventing rain-water coming into the room.

Another regular patron of the Farash Hotel is the beautiful golden-backed woodpecker (*Brachypternus aurantius*). This bird usually arrives later in the season than the coppersmith, but, like it, disdains a room which has been occupied by others. It is not, as a rule, so industrious as the coppersmith, for it usually selects for the site of its abode a part of the tree that is more or less hollow, and proceeds, by means of its pick-like beak, to cut out a neat round passage or tube leading to the ready-made cavity.

The most flashy of the *habitués* of the hotel is the Indian roller (*Coracias indica*), or " blue jay," as he is more commonly called. Like " loud " human beings, the roller bird is excessively noisy. When there are both green parrots and blue jays in the hotel it becomes a veritable bear-garden, resembling the hotels in Douglas, a town of the Isle of Man. During the summer months these are filled with holiday-makers from the Lancashire mills, who seem to spend the greater part of the night in playing hide-and-seek, hunt the slipper, " chase me," and such-like delectable games in the corridors and public rooms. There is, however, this difference between the rowdiness of the Lancashire " tripper " and that of the parrots and " jays "—the former is chiefly nocturnal, whereas the latter is strictly diurnal. The blue jays indulge in their screech-ings and caterwaulings, their aerial gymnastics, their " tricks i' the air," only during the hours of daylight. Not that the hotel is quiet at night. Far from it. The spotted owlets take care of that. The blue jay is not particular as to the nature of his accommodation ; any kind of hole is accepted, provided it be fairly roomy. He is quite content with a depression in the broken stump of an upright bough. Sometimes the bird places in its quarters a little furniture, in the shape of a lining of feathers, grass, and paper. More often the bird scorns such luxuries, and is content with the hard bare wood.

When a pair of blue jays first takes up its quarters in the hotel a great secret is made of the fact. Anyone who did not know the birds might think they were trying to avoid their creditors. This is not the case. The fact is that the nest contains some eggs which the owners imagine every other creature wants to steal. When, however, the young ones hatch out, the parents forget all about the necessity for concealing the whereabouts of the nest, so taken up are they with the feeding of their young ones.

The hoopoe (*Upupa indica*) is another bird that must be numbered among the *clientèle* of the hotel. It is just the kind of visitor that a hotel proprietor likes. It is not in the least particular as to its quarters. Any tumble-down room will do, the filthier the better! All that it demands is that the front-door shall be a mere chink, only just large enough to admit of its slender body. It then feels that its house is its castle; no enemy can possibly enter it.

The common myna (*Acridotheres tristis*) is another bird which habitually patronises the Farash Hotel. It is even less particular than the hoopoe as to the nature of its quarters—anything in the shape of a hole does quite well. Having secured accommodation, it proceeds to throw into it, pell-mell, a medley of straws, sticks, rags, bits of paper. That is its idea of house-furnishing. So untidy is the myna that you can sometimes discover the room it occupies by the pieces of furniture that stick out of the window! The mynas arrive later than most of the birds which nest in the farash, hence they find all the more desirable suites occupied. This does not distress the happy-go-lucky creatures in the least. They are probably the most contented of all the members of the little colony that lives in the *Hôtel des Oiseaux*. *Summæ opes, inopia cupiditatum.*

XIV

THE COOT

THE coot (*Fulica atra*) is a rail which has taken thoroughly to the water. It has, in consequence, assumed many of the characteristics of a duck. We may perhaps speak of it as a pseudo-duck. Certain it is that inexperienced sportsmen frequently shoot and eat coots under the impression that they are " black duck." Nevertheless, there is no bird easier to identify than our friend, the bald coot. In the hand it is quite impossible to mistake it for a duck. Its toes are not joined together by webs, but are separated and furnished with lobes which assist it in swimming. Its beak is totally different from that of the true ducks. But there is no necessity to shoot the coot in order to identify it. Save for the conspicuous white bill, and the white shield on the front of the head, which constitutes its " baldness," the coot is as black as the proverbial nigger-boy. Thus its colouring suffices to differentiate it from any of the ducks that visit India. Further, as " Eha " truly says, " its dumpy figure and very short tail seem to distinguish it, even before one gets near enough to make out its uniform black colour and conspicuous white bill." The difficulty which the coot experiences in rising from the water is another easy way of identifying it. Ducks rise elegantly and easily ; the coot plunges and splashes and beats the water so vigorously with wings and feet that it appears to run along the surface for a few yards before it succeeds in maintaining itself in the air. But, when fairly started, it moves at a great pace, so that, as regards flight, it may well say, even at the risk of perpetrating a pun, *Il n'y a que le premier pas qui coute.* During the efforts preliminary to flight the bird presents a very easy mark ; hence its popularity among inexperienced sportsmen. Now, since the coot is, to use a racing term, so indifferent a starter, raptorial birds must find it a quarry particularly easy to catch. Therefore, according to the rules of the game of natural selection, as drawn up by the learned brotherhood of zoologists, the coot ought to be as difficult to see as a thief in the night, and should spend its life skulking among rushes, in order to escape its foes. As a matter of fact it is as conspicuous as a lifeguardsman in full uniform, and, so far from having the habits of a skulker, it seems to take a positive delight in exposing itself, for, as Jerdon says, " It is often seen in the middle of some large tank far away from weeds or cover."

Someone has suggested that the coot is an example of warning colouration, that it is unpalatable to birds of prey, and that its black livery and white face are nature's equivalent to the druggist's label bearing the legend " Poison." Unfortunately for this suggestion, certain sportsmen, as we have seen, never lose an opportunity of dining off roast coot, and appear to be none the worse for the repast. Moreover, Mr. Frank Finn, who holds that no man is properly acquainted with any species of bird until he has partaken of the flesh thereof, informs us that " coots are edible, but need skinning, as the skin is tough and rank in taste." Miss J. A. Owen has a higher opinion of the flavour of the bird. She maintains that coots are " very good for eating, but they are not often used for the table, chiefly because they are so difficult to pluck, except when quite warm." Further, low-caste Indians appear to be very partial to the flesh of our pseudo-duck. One of the drawbacks to water-fowl shooting in this country is the constant wail of the boatmen, " *Maro wo chiriya, sahib, ham log khate hain* " (Shoot that bird, sir, we people eat it). Neither expostulations nor threats will stay the clamour. The sportsman will enjoy no peace until he sacrifices a coot. If, then, human beings of various sorts and conditions can and do eat the coot, it is absurd to suppose that the creature is unpalatable to birds of prey, some of which will devour even the crow. It is true that I do not remember ever having seen an eagle take a coot, but how few of us ever do see raptorial creatures seize their victims ? What is more to the point, some observers have seen coots attacked by birds of prey. We are, therefore, compelled to regard the bald coot as a ribald fellow, who makes merry at the expense of modern zoologists by setting at naught the theory of natural selection as it has been developed of late.

Some may, perhaps, accuse me of never missing an opportunity to cast a stone at this hypothesis. To the charge I must plead guilty ; but at the same time I urge the plea of justification. The amount of non-sense talked by some naturalists in the name of natural selection is appalling. The generally accepted con-ception of the nature of the struggle for existence needs modification. Natural selection has of late become a kind of fetish in England. So long as biologists are content to fall down and worship the golden calf they have manufactured, it is hopeless to look for rapid scientific progress. The aspersions I cast on Wallaceism are either justified or they are not. If they are justified, it is surely high time to abandon the doctrine of the all-sufficiency of natural selection to account for the whole of organic evolution. If, on the other hand, they are not justified, why do not the orthodox biologists arise and refute my statements and arguments ? It is my belief that the black livery of the coot is not only not the product of natural selection, but is positively harmful to its possessor ; that the coot would be an even more successful species than it now is, if, while retaining all its habits and other characteristics, it had a coat of less conspicuous hue. I maintain that many organisms possess characters which are positively injurious to them, and yet manage to survive. Natural selection has to take animals and plants as it finds them—their good qualities with the bad. If a species comes up to a certain standard, that species will be permitted to survive, in spite of some defects. By the ill-luck of variation the coot has

G

acquired black plumage, but this ill-luck is out-
weighed by its good-luck in possessing some favourable
characters.

The first of these favourable attributes is a good
constitution. Thanks to this the coot is able to thrive
in every kind of climate: in foggy, damp England;
in the hot, steamy swamps of Sind, and in cold Kashmir. In this respect it enjoys a considerable advantage
over the ducks, inasmuch as it is not exposed to the
dangers and tribulations of the long migratory flight.

Another valuable asset of the coot is a good digestion.
Creatures which can live on a mixed diet usually do
well in the struggle for existence. Then, the coot is a
prolific bird. It brings up several broods in the year,
and its clutch of eggs is a large one. The nest is usually
well concealed among reeds and floats on the surface
of the water, so is difficult of access to both birds and
beasts of prey. Moreover, the mother coot carefully
covers up the eggs when she leaves the nest. Another
useful characteristic of the coot is its wariness. Many
water-fowl go to sleep in the daytime, but the coot
appears to be always watchful. This perhaps explains
its popularity with ducks and other water birds,
although I should be inclined to attribute it to the
extreme amiability of the coot. Nothing seems to
ruffle him, except the approach of a strange male bird
to the nest. Whatever be the reason therefor, the
general popularity of the coot among his fellow-water-
fowl is so well established that in England many
sportsmen encourage coot on to their waters in order
to attract other water-fowl. Thus, a strong con-

stitution, a good digestion, prolificness, and wariness, enable the coot to thrive, in spite of its showy livery. The first three of the above characteristics enable the species to contend successfully with climate and disease, which are checks on the increase of organisms far more potent than predaceous animals. It is also possible—but this has yet to be domonstrated— that the coot, although edible, is not considered a delicacy by birds of prey, and so is taken when nothing more dainty is obtainable. If this be the case, it could, of course, minimise the disadvantages of the coot's conspicuousness. But even then there is no evading the fact that the blackness of the coot is an un-favourable characteristic.

XV

THE BEAUTIFUL PORPHYRIO

THE bald coot is, as we have seen, a rail that has taken thoroughly to an aquatic life. The purple coot may be described as a rail, which, while displaying hankerings after a life on the liquid element, has not definitely committed itself to the water. The porphyrio, then, is a rail which, to use a political expression, is "sitting on the fence." The indecision of Mr. Porphyrio has somewhat puzzled ornithologists. These seem to be unable to come to an agreement as to what to call him. Jerdon dubs him a coot, Blanford a moor-hen. The New Zealanders term him a swamp-hen, and their name is better than that given him by either Jerdon or Blanford, as denoting that the bird is neither a coot nor a moor-hen. But, perhaps, the classical name best suits a bird which is arrayed in purple and fine linen. For the fine linen, please look under the tail, at what Dr. Wallace would call the bird's recognition mark, although I am sure it will puzzle that great biologist to say what use so uniquely plumaged a bird as the porphyrio can have for a recognition mark. As well might Napoleon have worn a red necktie, to enable his friends to recognise him when they met him ! But this is a digression.

The Greeks were well acquainted with a near relative of the Indian porphyrio, which they kept in confinement. " For a wonder," writes Finn, " they did not keep it to eat, but because they credited it with a strong aversion to breaches of the conjugal tie in its owner's household." He adds : " Considering the state of morality among the wealthier Romans, I fear that accidents must often have happened to pet porphyrios."

The purple moor-hen is a study in shades of art blue —a bird which should appeal strongly to Messrs. Liberty and Co. Its bill, which is not flat like that of a duck, but rounded, is bright red, as is the large triangular shield on the forehead. Its long legs and toes are a paler red. The plumage is thus described by Blanford : " Head pale, brownish grey, tinged with cobalt on cheeks and throat, and passing on the nape into the deep purplish lilac of the hind neck, back, rump, and upper tail-coverts ; wings outside, scapulars and breast light greenish blue ; abdomen and flanks like the back ; the wing and tail-coverts black, blue on the exposed portions ; under tail-coverts white."

So striking a bird is this coot, that it cannot fail to arrest one's attention. Many sportsmen seem unable to resist the temptation of shooting it. Mr. Edgar Thurston informs me that a cold weather never passes without some sportsman sending him a specimen of *Porphyrio poliocephalus* for the Madras Museum. They come across the bird when out snipe-shooting, and, thinking it a rare and valuable species, pay it the very doubtful compliment of shooting it. As the museum has now a sufficient stock of stuffed porphyrios to meet its requirements for the next few decades, I hope that sportsmen in that part of the world will in future stay their hand when they come across the beautiful swamp-hen.

Rush-covered marshes, lakes, and *jhils*, which are overgrown with reeds and thick sedges, form the happy hunting-grounds of this species. Its long toes enable it to run about on the broad floating leaves of aquatic plants. They also make it possible for the bird to cling to the stems of reeds and bushes. Very strange is the sight it presents when so doing—a bird as big as a fowl behaving like a reed warbler. The long toes of the porphyrio are not webbed, but are provided with narrow lobes which enable it to swim, though not with the same ease as its cousin, the bald coot.

In places where it is abundant the purple swamp-hen is very sociable, and keeps much more to cover than does the coot. When flushed, it flies well and swiftly, with its legs pointing backwards—the position so characteristic of the legs of the heron during flight. Its diet is largely vegetarian, and it is said to commit much havoc in paddy fields. The harm it does is probably exaggerated, for the purple coot flourishes in many districts where no paddy is grown.

This species has one very unrail-like habit, that of taking up its food in its claws. Its European cousin, *P. veterum*, was seen by Canon Tristram " to seize a duckling in its large foot, crush its head and eat its brains, leaving the rest untouched." This behaviour Legge stigmatises as cannibalism ! There is no evidence that the purple moor-hen is a cannibal, but it is not safe to keep the bird in the same enclosure as weaker birds.

Its voice is not melodious ; indeed, it is scarcely more pleasant to refined ears than the wail of the street-singer.

Purple coots breed in company. The nest is a platform made of reeds and rushes, or, when these are not available, of young paddy plants, erected on a tussock of long grass projecting out of the water, usually some way from the edge of the *jhil*. Hume's observations led him to lay down two propositions regarding the nesting habits of this species. First, " that all birds in the same swamp both lay and hatch off about the same time." Secondly, " that in two different *jhils* only a dozen miles apart, and, apparently, precisely similarly situated, there will be a difference of fifteen days or more in the period of the laying of the two colonies." Neither of these statements appears to hold good of the purple coot in Ceylon, for, according to Mr. H. Parker, " they do not breed there simultaneously." " Young birds, eggs in all stages of incubation and partly built nests are all found in the same tank. In some cases the eggs are laid at considerable intervals. I have met with a nestling, partly incubated eggs of different ages and fresh eggs in the same nest." Widely distributed species not infrequently display local variations in habit. Such local peculiarities are of considerable interest, for they must some-times form the starting-points for new species. They are also responsible for some of the discrepancies which occur in the accounts of the species by various ob-servers. The nesting season is from June to September ; August for choice, in India. The eggs are pale pink, heavily splashed with red, quite in keeping with the beautiful plumage that characterises the adult bird. Sometimes the eggs of purple coots are placed under the barn-door fowl. Young porphyrios hatched under such conditions become quite tame and form a pleasing addition to the farmyard.

XVI

THE COBRA

ACCORDING to my dictionary, the cobra di capello (*Naia tripudians*) is a reptile of the most venomous nature. This, like many other things the dictionary says, is not strictly true. There exist snakes whose bite is far more poisonous than that of the cobra. The common krait, for example, is four times as venomous, and yet the bite of this little reptile is mild as compared with that of the sea snake, which should be carefully distinguished from the sea-serpent of the " silly season." But let us not quarrel with the writer of the dictionary ; he did his best. The cobra is quite venomous enough for all practical purposes to merit the title of " the most venomous." A fair bite kills a dog in from five minutes to an hour. Notwithstanding the lethal nature of his bite, the cobra is said by all who know him intimately to be a gentle, timid creature. Sulkiness is his worst vice. In captivity he sometimes sulks to such a degree as to starve to death unless food be pushed down his gullet ! The cobra is a reptile who prefers retiring gracefully to facing the foe. It is only when driven into a corner that he strikes, and then

apparently he does so with the utmost reluctance.
Nicholson writes : " A cobra standing at bay can be
readily captured ; put the end of a stick gently across
his head and bear it down to the ground by a firm
and gradual pressure. He will not resist. Then place
the stick horizontally across his neck and take him
up. You must not dawdle about this ; sharp is the
word, when dealing with snakes, and they have as
much respect for firm and kind treatment as contempt
for timidity and irresolution." " There is very little
danger," he adds, " about handling this snake ; nerve
is all that is required." I have no doubt that this
is all true. It is certainly borne out by the non-
chalance with which an Indian, who is accustomed
to snakes, will put his hand into a basket of cobras and
pull one out. There are, however, some things the
doing of which I prefer to leave to others, and one of
these is the handling of venomous snakes. There is
always the colubrine equivalent of the personal equa-
tion to be taken into consideration. People whose
fondness for playing with fire takes the form of snake-
charming will do well to operate upon light-coloured
specimens, for experience has taught those who handle
snakes that dark-coloured varieties are worse-tempered
than those of paler hue. In some unaccountable
manner blackness seems to be correlated with evil
temper. Another word of warning. A snake has a
longer reach than might be anticipated. On one
occasion, wishing to show how the cobra strikes,
I walked up to within a yard or two of one standing at bay and threw a clod of earth at him. He

struck, and his head came unpleasantly near to my legs !

The cobra is a species of considerable interest to the zoologist. In the first place, several varieties exist. Some cobras have no figure marked on the hood, others display a pattern like a pair of spectacles, while others show a monocle. These are known respectively as the anocellate, the binocellate, and the monocellate varieties. The binocellate form is most frequently met with. It is found all over India. It is the only variety that occurs in Madras, and the one most commonly found in Bombay and North-Western India. The great majority of the cobras that dwell in Central India belong to the anocellate variety. This form is also found on the frontier from Afghanistan to Sikkim. The monocellate variety is the common cobra of Bengal, Burma, and China.

There can be but little doubt that the cobra is a form undergoing active evolution. *Naia tripudians* appears to be splitting up into three species. The spectacled cobra is probably the ancestral form. The black anocellate variety seems best adapted to the climatic conditions of the Central Provinces, while the pale, binocellate form thrives in Southern India. It is possible that these external characteristics are in some way correlated with adaptability to particular environments. Curiously enough, brown, yellow, and black varieties of the African cobra (*Naia haje*) exist. Some species of birds display a similar phenomenon. The coucal or crow-pheasant, for example, is divided up into three local races. Most naturalists are agreed

that geographical isolation has been an important
factor in the making of some species. Exactly why
this should be so has yet to be explained.

Another interesting feature of the genus *Naia* is the
dilatable neck or hood. Of what use is this to its possessor ? Zoologists, or at least those of them who sit
at home in easy chairs and formulate theories, have
an answer to this question. They assert that the hood
has a protective value. A cobra when at bay raises
the anterior portion of its body, expands its hood, and
hisses. This is supposed to terrify those animals which
witness the demonstration. Thus Professor Poulton
writes : " The cobra warns an intruder chiefly by
attitude and the broadening of its flattened neck, the
effect being heightened in some species by the ' spectacle.' " Unfortunately for this hypothesis, no creature, with the possible exception of man, appears to be
in the least alarmed at this display. Dogs regard it as
a huge joke. Of this I have satisfied myself again and
again, for when out coursing at Muttra we frequently
came across cobras, which the dogs used invariably
to chase, and we sometimes found it very difficult to
keep the dogs off, since they seemed to be unaware
that the creature was venomous. Colonel Cunningham's experience has been similar. He writes :
" Sporting dogs are very apt to come to grief where
cobras abound, as there is something very alluring to
them in the sight of a large snake when it sits up
nodding and snarling ; and it is often difficult to come
up in time to prevent the occurrence of irreparable
mischief." He also states that many ruminants have

a great animosity to snakes and are prone to attack any that they may come across. We must further bear in mind that even if the cobra does bite his adversary, this will avail him nothing, for the bite itself, though painful, is not sufficiently so to put a large animal *hors de combat* immediately. It does not profit the cobra greatly that his adversary dies after having killed him.

Thus, it seems to me that neither the hood nor the venom is protective. Indeed, it is difficult to understand how it is that the poison fangs have been evolved. The venom, of course, soon renders a small victim quiescent and so makes the swallowing of it easier than would otherwise be the case. But non-venomous snakes experience no difficulty in swallowing their prey. Moreover, in order that natural selection can explain the genesis and perfecting of an organ it is not sufficient to show that the perfected organ is of use. We must demonstrate that from its earliest beginning the organ in question has all along given its possessor sufficient advantage in the struggle for existence to effect his preservation when his fellows have been killed.

XVII

THE MUNGOOSE

FROM the cobra it is a natural step to his foe—the mungoose. This creature—the ichneumon of the ancients—occupies a most important place in the classical and mediæval bestiaries. Every old writer gives a graphic account, with variations according to taste, of the " mortall combat " between the aspis and the ichneumon. But the noble creature was not content with fighting a mere serpent, it used to pit itself against the leviathan. Pliny tells us that the crocodile, having gorged himself, falls asleep with open mouth in order that the little crocodile bird may enter and pick his teeth. Then the watchful ichneumon "whippeth " into the monster's mouth and "shooteth" himself down his throat as quick as an arrow. When comfortably inside, the ichneumon sups off the bowels of his host, and, having satisfied his hunger, eats his way out through the crocodile's belly, so that, to use the words of the learned Topsell, who gallantly gives *place aux dames*, " Shee that crept in by stealth at the mouth, like a puny thief, cometh out at the belly like a conqueror, through a passage opened by her own labour and industrie."

In these degenerate days the mungoose does not perform such venturesome exploits ; nevertheless, he still has a " bold and sanguinary disposition." Sterndale's tame mungoose once attacked a greyhound. Although in the wild state he does nothing so quixotic as to assail large snakes, the mungoose is a match for the cobra. The natives of India declare that, when bitten by his adversary, he trots off into the jungle and there finds a root or plant which acts as an antidote to poison, so that he may claim to be the discoverer of the anti-venom treatment for snake-bite. We may term this the anti-venom theory to account for the immunity of the mungoose. It bears the stamp of antiquity, but is unsupported by any evidence. In this respect it is not much worse off than some modern zoological theories. The other hypothesis we may call the-prevention-is-better-than-cure theory. It attributes the immunity of the mungoose to his remarkable agility. He does not allow the cobra to " have a bite," and even if the latter does succeed in striking, the chances are that its fangs will be turned aside by the erected hair of the mungoose or fail to penetrate his tough skin. Blanford states that although it has been repeatedly proved that the little mammal dies if properly bitten by a venomous snake, it is less susceptible to poison than other animals. He adds : " I have seen a mungoose eat up the head and poison glands of a large cobra, so the poison must be harmless to the mucous membrane of the former animal."

Eight species of mungoose occur in the Indian

Empire. The only one which is well known is the common mungoose, which Jerdon calls *Herpestes griseus*. It is, I believe, now known as *Herpestes mungo*. During the last century it has been renamed some eight or nine times.

It is not necessary to describe the mungoose. The few Anglo-Indians who have not met him in the wild state must have frequently seen him among the " properties " of the individual who calls himself a snake-charmer.

The mungoose lives in a hole excavated by itself. It is diurnal in habits, and feeds largely on animal food. Jerdon states that it is " very destructive to such birds as frequent the ground. Not infrequently it gets access to tame pigeons, rabbits, or poultry, and commits great havoc. . . . I have often seen it make a dash into a verandah where some cages of mynas, parrakeets, etc., were daily placed, and endeavour to tear them from their cage." But birds are not easy for a terrestrial creature to procure, so that its animal food consists chiefly of mice, small snakes, lizards, and insects. Jerdon states that " it hunts for and devours the eggs of partridges, quail, and other ground-laying birds." I am inclined to think that the carnivorous propensities of the mungoose have been exaggerated, for its food seems to contain a considerable admixture of vegetable substances. In captivity it will eat bread and bananas, although it requires animal food in addition. McMaster records the case of a mungoose killed near Secunderabad, of which the stomach contained a quail, a portion of a custard apple, a small wasp's nest, a blood-sucker lizard, and a number of insects—quite a *recherché* little repast !

In Lahore I, or rather my wife, made the discovery that the mungoose is very fond of bird-seed. A certain individual contrived to spend the greater part of the day in our bungalow. He was probably attracted in the first instance by the amadavats. Finding that these were secure in their strongly-made cage, he turned his attention to their seed, and found that it was good. When he had devoured all that had fallen to the ground he would endeavour by means of his claws to extract seed from within the cage. This used to alarm the birds terribly ; one night their flutterings woke me up. It takes an amadavat a long time to learn that it is safe in its cage. It is not until after months of captivity that it will sit on the floor of its house and gaze placidly at the hungry shikra which has alighted on the top. For this reason we did not encourage that mungoose. I may say that we distinctly discouraged it by throwing things at it, or chasing it out of the bungalow whenever we saw it. But it soon became so bold that, unless we ran out of the bungalow after it, it used to remain in hiding in the verandah, and, a few seconds after all was quiet, its little nose would appear at the doorway.

The impudence of the Indian house-crow is great, that of the sparrow is colossal, that of the striped squirrel staggering, but the impudence of all these is surpassed by that of the mungoose. Small wonder, then, that it makes an excellent pet. McMaster kept one that died of grief when separated from him. But,

H

in order to tame a mungoose, the animal must be captured while young. Babu R. P. Sanyal, in his useful *Handbook on the Management of Animals in Captivity*, writes : " Adult specimens seldom become tame enough even for exhibition in a menagerie ; they either remain hidden away in the straw or snap at the wire, uttering a querulous yelp, possibly expressive of disgust, at the approach of man. They have been known to refuse nourishment and to starve to death."

A mungoose (*Herpestes ichneumon*) allied to our Indian species is common in Egypt, where it is known as Pharaoh's rat or Pharoe's mouse. It is frequently trained by the inhabitants to protect them from rats and snakes.

The mungoose is a ratter without peer. Bennet, in his *Tower Menagerie*, states that " the individual now in the Tower actually, on one occasion, killed no fewer than a dozen full-grown rats, which were loosed to it in a room 16 feet square, in less than a minute and a half." The Egyptian species eats crocodiles' eggs, so that Diodorus Siculus remarks that but for the ichneumon there would have been no sailing on the Nile. The Indian species seems to display no penchant towards crocodiles' eggs.

XVIII

THE SWAN

"With that I saw two swannes of goodly hewe
Come softly swimming downe the lee;
Two fairer birds I yet did never see;
The snow, which does the top of Pindus strew,
Did never whiter shew."

WHEN I speak of " the swan," I mean the
bird called by ornithologists the mute
swan (*Cygnus olor*), the swan of the poets
that warbles sublime and enchanting
music when it is about to shuffle off its mortal coil, the
tame swan of Europe, the swan that used to take
Siegfried for cheap trips down the river, the swan that
" graces the brook," the swan of the " stately homes
of England," the swan I used to feed as a youngster
on the Serpentine, not the black fellow in St. James's
Park, the swan that hovers expectantly in the offing
while you are having tea in a boat on the Thames.
This is, of course, by no means the only species of swan.
There are plenty of others—white ones, black ones,
black-and-white ones—for the family enjoys a wide
distribution. Nevertheless, I propose to confine
myself to this particular swan. I have excellent
reasons for doing so. As it is the only swan with which
I have had much to do, I can, like the Cambridge Don

99

who declared that the Kaiser was quite the pleasantest
Emperor he had ever met, say that *Cygnus olor* is the
most agreeable of my swan acquaintances. This may
sound like flattery, like the fulsome praise of the
penny-a-line puffer. It is nothing of the kind. It is
barely complimentary. Among the blind the one-eyed
is king, unless, of course, he lives in a republic. " You
are the best of a very bad lot," were the encouraging
words with which a prize for arithmetic was once
handed to me. The mute swan is the most agreeable
of a bad-tempered clan.

Swans are overrated birds. They cannot hold a
candle to their despised cousins, the geese. I am sorry
to have to say this, to thus shatter another idol of the
poets, to expose yet another of what the Babu would
call their " bull cock " stories. I am the more sorry
as I am fully aware that this will bring down upon me
the thunderous wrath of the literary critic, whose
devotion to the British bards is truly affecting. Let me,
therefore, by way of trimming, say that there is some
justification for idolising the swan. The bird is as
beautiful as the heroine in a three-volume novel. He
is dignified and stately, full of " placid beauty."
" Proudly and slow he swims through the lake in the
evening stillness. No leaf, no wave, is moving : the
swan alone goes on his solitary course, floating silently
like a bright water spirit. How dazzlingly his snowy
whiteness shines ! How majestically the undulating
neck rises and bends ! With what lightness and
freedom he glides buoyantly away, the pinions unfurled
like sails ! Each outline melts into the other ; every

attitude is full of feeling, in every movement is nobility : an ever-changing play of graceful lines, as though he knew that the very stream tarried to contemplate his beauty."

But his splendour is not without alloy. It is marred by the tiny, black, beady eye, which gives the bird an evil-tempered, sinister expression. This expression is in keeping with the character of the swan. Cygnus is a bully. He delights to tyrannise over the ducks who so often keep him company in captivity. The domineering behaviour of an old swan that used to live in the Zoological Gardens at Lahore was amusing to watch. The water-fowl are fed twice daily, the food being placed in a series of dishes so that all can eat at once. The swan used to appropriate the first dish to be filled, and no duck or goose durst approach that dish. Having taken the edge off his appetite, the swan would waddle to the next plate, and drive away the ducks that were eating out of it. He would then pass on to dish number three, and so on all along the line, his idea being, apparently, to cause the maximum of annoyance to his neighbours with the minimum of trouble to himself. There were great rejoicings among the ducks when that old swan died.

An angry swan is capable of inflicting a nasty blow with its powerful wings. It is said to be able to use these with such force as to break a man's arm. Mr. Kay Robinson denies this, and declares that the wing of a swan is not a formidable weapon. Personally, I always give the wing the benefit of the doubt and an angry swan a wide berth.

Considering its size, the swan has a very small brain ; hence it is not overburdened with intelligence. Mr. H. E. Watson relates how one day when shooting in Sind he came across five swans on a tank. " They let the boat get pretty close," he writes, " and I shot one. The other four flew round the tank a few times and then settled on it again. I went up in the boat and fired again, but without effect. They flew round, and then settled again. The third time I shot another ; the three remaining again flew round and settled, and the fourth time I fired I did not kill. Exactly the same thing happened the fifth time ; the birds flew round and round, and settled close to me, and I shot a third. The remaining two flew a little distance, and settled, but I thought it would be a pity to kill them . . . so I began to shoot ducks, and then the two remaining swans flew by me, one on the right and one on the left, so that I could easily have knocked them over with small shot." What a pity swans are such rare visitors to India ! What grand birds they must be for an indifferent shot. One swan on a small *jhil* would give a really bad gunner a whole morning's shooting ; it would circle round and round the sportsman at short range, letting him blaze off to his heart's content until it fell a victim to its trustfulness ! Try to imagine the so-called stupid goose behaving in this manner.

The swan is a very silent bird in captivity, for this reason it is called the mute swan. The only noise I have ever heard it make is a hiss when it is angry. At the breeding season it is said to trumpet sometimes. The ancients believed that the swan, though mute

throughout life, sings most sublimely at the approach
of death ; it then sings, not a funeral dirge, but a
jolly, rollicking song. This presented an excellent
opportunity for moralising. Mediæval authors were
always on the look out for such opportunities. The
swan, wrote the author of the *Speculum Mundi*, " is
a perfect emblem and pattern to us, that our death
ought to be cheerful, and life not so dear to us as it is."
This practice of singing before death has, like the crino-
line, quite gone out of fashion. The mute swan can
never have been so great a musician as some of his
brethren, since the French horn which he carries in his
breast-bone is not nearly so well developed as it is in
either the hooper or the black swan. Let me here
say, *en passant*, that both ancient and mediæval writers
declined to believe in the existence of a black swan ;
they regarded it as " the very emblem and type of
extravagant impossibility." The phœnix, the dragon,
and the mermaid they could believe, but they felt that
they must really draw the line at a black swan.

A swan's nest is a bulky structure composed of
rushes, reeds, and other aquatic plants ; it is placed
on the ground near the water's edge. Six or seven
large greenish-white eggs are laid. The breeding season
is from March to May. Swans do not, of course, breed
in India. Indeed, it is only on rare occasions that
they visit that country, and then they do not venture
farther south than Sind.

XIX

KITES OF THE SEA

"Graceful seagulls, plumed in snowy white,
Follow'd the creaming furrow of the prow,
With easy pinion, pleasurably slow ;
Then on the waters floated like a fleet
Of tiny vessels, argosies complete,
Such as brave Gulliver, deep wading, drew
Victorious from the forts of Blefuscu."

OF all the methods of obtaining food to which
birds resort, none makes greater demands
on their physical powers than that which
we human beings term scavenging—the
seeking-out and devouring of the multifarious edible
objects left, unclaimed by the owners, on the face of the
land or the sea. No bird can eke out an existence by
scavenging unless it be endowed with wonderful power
of flight, the keenest eyesight, and limitless energy,
to say nothing of the ability and the will to fight when
necessity arises. Thus it happens that it is to the
despised scavengers that we must direct our eyes
if we would behold the perfection of flight. The
vultures, the kites, and the gulls are verily the monarchs
of the atmosphere.

Bird scavengers are of two kinds—specialists and
general practitioners. The former confine themselves

to one particular kind of food—the bodies of dead animals. Of such are the vultures. In the polity of the feathered folk might is right, so that these great birds enjoy the prerogative of picking and choosing their food. The lesser fry have to be content with that which the vultures do not require, with the crumbs that fall from the vulturine table ; they are ready to devour " anything that is going." All is grist that comes to their mill.

The kites and gulls are the chieftains of the clan of general scavengers. The sway of the former extends over the land : the latter have dominion over the seas. Kites cannot swim ; their operations are in consequence necessarily confined to the land, and to water in the neighbourhood of *terra firma*. Sea-gulls, on the other hand, are as buoyant as corks, and have webbed feet ; they are, further, no mean swimmers, and are eminently adapted to a seafaring life. They are birds of powerful flight, and almost as much at home on land as at sea. They confine their attention mainly to the sea, not because they are compelled by their structure to do so, but because they encounter less opposition there.

Among birds, similarity in feeding habits often engenders similarity in appearance—a professional likeness grows up among those that pursue the same calling. The likeness between swifts and swallows is a remarkable instance of this. The separate sphere of influence occupied by kites and gulls sufficiently explains the dissimilitude of their plumage. In nearly all other respects the birds closely resemble one another.

In habits, gulls are marine kites. Grandeur of flight
is the most marked attribute of each. They do not
cleave the air at great velocity, like swifts or " green
parrots." It is the effortlessness, the perfect ease with
which kites and sea-gulls perform their aerial movements for hours at a time, rather than phenomenal
speed, that compels our admiration. A dozen gentle
flaps of the wings in a minute suffices to enable a gull
to keep pace with a fast steamship.

Cowper sang of—

> " Kites that swim sublime
> In the still repeated circles, screaming loud."

These words are equally appropriate to the kites of
the sea.

I have watched, until my eyes grew tired, kites
floating in circles in the thin atmosphere, with scarce
a movement of the pinions ; I have seen gulls keeping
pace with a steamer without as much as a quiver of
their wings. In each case the wind was the motive
power.

Both kites and gulls fly with downwardly directed
eyes. Their life is one long search for food. So keen
is their vision that no object seems minute enough to
escape their notice. The smallest piece of bread
thrown from a moving ship is immediately pounced
upon by the " wild sea-birds that follow through the
air," but no notice appears to be taken of a piece of
paper rolled up into a ball.

Gulls, like kites, are omnivorous. Some species
occasionally prey upon fish which they catch alive ; this
method of obtaining food is, however, the exception

rather than the rule among gulls. They are sea-birds merely in the sense that they usually pick their food off water. They are found only where there is refuse to be picked up. In those parts of the ocean that are not frequented by ships gulls are conspicuous by their absence. They do not, as a rule, travel very far from land ; when they do venture out to sea, it is invariably in the wake of some great ship. Every ocean liner sheds edible objects all along its course, and so attracts numbers of gulls. These follow the ship for perhaps two hundred miles, and then forsake it to return with some homeward-bound vessel.

The seashore and the estuaries of tidal rivers are the favourite hunting-grounds of the sea-gulls, the flotsam of the rivers and the jetsam of the waves being the attractions. Numbers of the graceful birds await the return of the fishing smacks, in order to secure the fish thrown away by the fishermen. The marine kites are not always content to wait for rejected fish ; not infrequently they boldly help themselves to some of the shining contents of the nets, and sometimes actually tear the meshes with their strong sharp bills. In India there is always much fighting between the gulls and the crows over the fish cast away by the fishers. The antagonists are well matched. Similar contests have been recorded in the British Isles. I cull from *The Evening Telegraph* the following description of a fight between gulls and rooks over ground covered with worms which had been killed by a salt-water flood : " Thousands of gulls and rooks fought each other with a determination and venom that could

not be appreciated unless witnessed. Feathers flew
in all directions ; the cawing and screaming were
almost deafening. It was a genuine fight. At first it
took place in mid-air, but soon the combatants came
to the ground, and then the struggle centred in and
around a fairly large hillock. Just as the gulls appeared
to be gaining the upper hand, the report of a gun broke
up the fight."

The diet of the kites of the sea is not confined to
small things. " A son of the marshes " states that
he has seen them feeding with hooded crows on the
carcases of moorland sheep. In the British Isles gulls
frequently follow the plough and greedily seize the
worms and grubs turned up in the furrow. In London
and Dublin, and probably in other places, gulls have
taken up their residence in the parks, where they feed
largely on the bread thrown to the ornamental water-
fowl, seizing it in the air before it reaches the ducks.
So tame do these gulls become that they will almost
take bread from the hands of children. Many people
labour under the delusion that these gulls are domesticated ones kept by the authorities along with the
ducks and swans.

Of late years a large colony of gulls has established
itself on the Thames opposite the Temple. These now
form one of the sights of London. The townsfolk take
so much interest in the graceful birds that some individuals earn a living by selling on the Embankment
small baskets of little fish which passers-by purchase
in order to throw to the screaming gulls that hover
around expectantly.

Even as hunger frequently drives kites to commit larceny in the farmyard, so does it sometimes turn sea-gulls into birds of prey.

Mr. W. J. Williams gives an account, in *The Irish Naturalist*, of a lesser black-headed gull that used to frequent the lake in St. Stephen's Green Park. It was wont to rest on the cornice of a house overlooking the park, till an opportunity presented itself of swooping down and snatching a duckling. It became so expert at this form of poaching that the Board of Works had the marauder executed. Another gull which attacked a duckling was in turn attacked by the parents (a pair of Chilian wigeons), with such success that the exhausted gull was killed with a stick by one of the Park constables.

In India gulls do not, I think, venture far inland. The terns regard the inland waters of Hindustan as their preserve. Some people eat gulls. The late Lord Lilford declared that the black-headed species is a good bird for the table. I am not prepared to deny this assertion. I shall not put it to the test, for, in my opinion, gulls should be a feast only for the eyes.

XX

RIVER TERNS

A SOJOURN of a few years in Upper India usually teaches a European to make the most of the cold weather as it gives place to the heat of summer. There is a period of a week or two in March and early April when, although the days are very hot, the nights and early mornings are cool, when the mercury in the thermometer fluctuates between 104° and 68° F. If at this season a man is energetic enough to rise at 5.15, shortly after the birds awake, there are few more pleasant ways of spending the ensuing three hours than by taking what the French would term a promenade upon the water. The gliding motion of a boat propelled by sail or oar is always soothing, and is doubly so when one knows that the breeze which then blows cool upon the cheek will scorch the face seven hours hence. The morning excursion on the water is rendered especially enjoyable if it happens to take place at one of the comparatively few parts of the Ganges or the Jumna where the river-bed is narrow, so that the water fills the space between the banks, instead of being, as is more usually the case, a mere trickle of water meandering through a great expanse of sand.

Under the former conditions it is good to sit in the stern of a gliding boat and watch the birds that frequent the river.

At sunrise the crow-pheasants (*Centropus rufipennis*) come to the water's edge to drink, so that numbers of the long-tailed, black birds with chestnut wings are to be seen from the boat. Having slaked their thirst, they hop up the steep bank with considerable dexterity, to disappear into the stunted bushes that grow on the top of the bank. Then there are, of course, the regular *habitués* of the water's edge—the birds that frequent it at all hours of the day—the ubiquitous paddy bird (*Ardeola grayii*), which spends the greater part of its life ankle-deep in water, waiting motionless for the coming of its prey ; the common sandpiper (*Totanus hypoleucus*), that solitary bird, as small as a starling, which, on the approach of a human being, emits a plaintive cry and flies away, displaying pointed wings along the length of which runs a narrow white bar ; the handsome spur-winged plover (*Hoplopterus ventralis*), whose call is very like that of the did-he-do-it—but we must not dwell on these littoral birds, for to-day I would write of terns, the river birds *par excellence*. None of God's creatures are more attractive than terns to those who love beauty. That few, if any, of our English poets have sung the praises of these beautiful birds surely demonstrates how little attention poets pay to nature, and how artificial are their writings. This will, I fear, annoy the friends of the poets. I am sorry, but I cannot help it. It is the fault of the bards for having so grossly neglected the terns.

In colouring, these superb birds show what endless possibilities are open to the artist who confines himself to black and white and their combinations.

There is in the flight of terns a poetry of motion over which no one with an eye for the beautiful can fail to wax enthusiastic. The popular name for terns—sea-swallows—is a tribute to their wing power. They are all designed upon a common plan. Length and slimness characterise every part of their anatomy, save the legs, which are very short. Terns rarely walk ; nearly all their movements are aerial.

The terns that commonly frequent the rivers of Upper India are of three species—the black-bellied tern (*Sterna melanogaster*), the Indian river tern (*S. seena*) with its deeply forked tail, and the whiskered tern (*Hydrochelidon hybrida*), a study in pale grey. These, when not resting on a sandbank, are dashing through the air without effort, ever and anon dropping on to the water to pick something from off the surface, or plunging in after a fish. Allied to the terns, and found along with them, are the Indian skimmers (*Rynchope albicollis*), easily recognised by their larger size and black wings.

The passing of a black crow causes some of the terns to desist from their piscatorial occupation, in order to mob the intruder. This means that there are terns' eggs or young ones in the vicinity. Many species of birds betray the presence of their nests by displaying unusual pugnacity at the breeding season. To discover the eggs or young of the terns is not a difficult matter. It is only necessary to land upon the nearest island between which and the river bank there is a sufficient depth of water to prevent jackals fording it. If the island contain eggs or young ones, the parent birds will make a hostile demonstration by collecting overhead and flying backwards and forwards, uttering their harsh cries, and the nearer one approaches the nest the more clamorous do they become. In this manner they unwittingly inform the nest-seeker whether he is getting " hot " or " cold," to use the expressions employed in a nursery game.

The terns which breed on islets in Indian rivers do not appear to do much incubating in the daytime. There is no need for them to do so, because the sand grows very warm under the rays of the sun. Moreover, the only foes to be feared are the crows and the kites, which the terns can keep at bay more effectually when on the wing than while sitting on the eggs. Very different is the behaviour of the sea terns, whose eggs are liable to attack by gulls and crabs. For safety's sake the sea terns lay in large colonies, and, to use Colonel Butler's expression, sit on their eggs " packed together as close as possible without, perhaps, actually touching one another." He once came upon the nests of a colony of large-crested terns (*Sterna bergii*). The sitting birds did not leave their eggs until he was within a few yards of them. Having put them up, he retired to a little distance. " No sooner had I done so," he writes, " than both species [i.e. the gulls and terns] began to descend in dozens to the spot where the eggs were lying. In a moment a general fight commenced, and it was at once evident that the eggs belonged to *Sterna bergii*, and that

I

the gulls were carrying them off and swallowing their contents as fast as they could devour them." River terns do not construct any nest. They deposit their eggs on the bare, dry sand. The eggs have a stone-coloured ground, sometimes suffused with pink, blotched with dark patches, those at the surface of the shell having a sepia hue, and those deeper down appearing dark greyish mauve. The eggs, although not conspicuous, may, without difficulty, be detected when lying on the sand. Their colouring would seem to be adapted to match a stony, rather than a sandy environment, but the fact that the colouring of the eggs is but imperfectly protective does not much matter when the latter lie on a sand island, to which but few predaceous creatures have access ; the watchfulness of the parent birds more than compensates for the comparative conspicuousness of the eggs.

Young terns, like most other birds, are born helpless, and are then covered with a greyish down ; but before the tail feathers have broken through their sheaths, and while the wing feathers are quite rudimentary, the ternlets learn to run about and swim upon the water. At this stage the little terns look like ducklings when on the water, and, as they run along the water's edge, may easily be mistaken, at a little distance, for sandpipers.

When a young tern is surprised by some enemy, his natural instinct is to crouch down, half buried in the sand, and to remain there quite motionless until the danger has passed. The colouring of his down is such as to cause him to assimilate more closely to the sandy environment than the eggs do. If one picks up such a crouching ternlet, the bird will probably not struggle at all ; it may, perhaps, peck at one's fingers, but in nine cases out of ten will remain limp and motionless in the hand, looking as though it were dead, and if it be set upon the ground it falls all of a heap, and remains motionless in the position it assumed when dropped. If you take a young tern in your hand and lay it upon its back on the sand it makes no attempt to right itself, but remains motionless in that attitude, looking for all the world like a trussed chicken ; but if you turn your back upon it, it will take to its little legs and run, with considerable speed, to the water, to which it takes just as a duck does, its feet being webbed at all stages of its existence.

XXI

GREEN BULBULS

SINCE green is a splendid protective colour for an arboreal creature, it is surprising that there are not more green animals in existence. The truth of the matter is that green seems to be a difficult colour to acquire. There does not exist a really green mammal ; while green birds are relatively few and far between. In India we have, it is true, the green parrots, the barbets, the green pigeons, the green bulbuls, and the bee-eaters. Take away these and you can count the remaining green birds on the fingers of your hands. Curiously enough, the bee-eaters spend very little time in trees ; consequently the beautiful leaf-green livery seems rather wasted on them. And of the other green birds we may almost say that they are precisely those that seem least in need of this form of protection. The parrakeets and barbets, thanks to their powerful beaks, are well adapted to fighting, while more pugnacious birds than bulbuls and pigeons do not exist. I think, therefore, that the green liveries of these birds are not the result of their necessity for protection from raptorial foes. This livery is a luxury rather than a necessity.

Anatomically speaking, green bulbuls are not bulbuls at all. Jerdon called them bulbuls because of their bulbul-like habits, although, as " Eha " points out, they take more after the orioles. Oates tells us that these beautiful birds are glorified babblers, rich relations of the disreputable-looking seven sisters. He gives them the name *Chloropsis*.

Seven species of green bulbul are found in India ; they thus furnish an excellent example of a bird dividing up into a number of local races. When the various portions of a species become separated from one another this phenomenon often occurs. The common grey parrot of Africa is, according to Sir Harry Johnston, even now splitting up into a number of local races. That interesting bird is presenting us with an example of evolution while you wait. It is quite likely that the process may continue until several distinct species are formed. We must bear in mind that there is no essential difference between a species and a race. When the differences between two birds are slight we speak of the latter as forming two races ; when the divergence becomes more marked we call them species. Very often systematists are divided as to whether two allied forms are separate species or mere races. In such cases some peacemakers split the difference and call them sub-species.

Green bulbuls are essentially arboreal birds. In the olden time when India was densely wooded I believe that there was but one species of *Chloropsis*, even as there is but one species of house-crow in India proper. Then, as the land began to be denuded of forest in parts, these green bulbuls became a number of isolated communities, with the result that they eventually evolved into several species. In this connection I may mention that the grey on the neck of *Corvus splendens* is much more marked in birds from the Punjab than in those that worry the inhabitants of Madras.

Of the green bulbuls only two species occur in South India—the Malabar *Chloropsis* (*C. malabarica*) and Jerdon's *Chloropsis* (*C. Jerdoni*). The former, as its name tells us, is found in Malabar. The green bulbul of the other parts of South India is Jerdon's form. This handsome bird does not occur in or about the City of Madras ; at least I have never seen it in the neighbourhood, nor indeed nearer than Yercaud. However, not improbably it occurs between the Shevaroys and the east coast. If anyone who reads these lines has seen this bird in that area, I hope that he or she will be kind enough to let me know. Here let me say that to identify a green bulbul is as easy as falling out of a tree. He is of the same size as the common bulbul. His prevailing hue is a rich bright grass-green—the green of grass at its best. His chin and throat are black, and he has a hyacinth-blue moustache, so that he deserves his Telugu name—the "Ornament of the Forest." His wife is pale green where he is black and her moustache is of a paler blue. The Malabar species is easily distinguished by its bright orange forehead. Green bulbuls go about, sometimes in small flocks, more frequently in pairs. They rarely, if ever, descend to the ground, but flit about amid the foliage, to which they assimilate so closely, seeking for the insects, fruit, and seed on which they feed. Like many other gaily attired birds, they give the lie to the oft-repeated assertion that it is only the dull-hued birds that are good songsters. Green bulbuls are veritable gramophones, " flagrant plagiarists " Mr. W. H. Hudson would call them. Not only have they a number of pretty notes of their own, but the feathered creature whose song they cannot imitate remains yet to be discovered. Green bulbuls might be called Indian mocking-birds were there not so many other birds in the country that imitate the calls of their fellows. Some ornithologist with a good ear for music should draw up a list of all our Indian birds that mock the calls of others, setting against each the names of these whose sounds they imitate.

Green bulbuls are hardy birds and thrive well in captivity. I saw recently a specimen in splendid condition at a bird show in London. " There is one drawback, however," writes Finn in his *Garden and Aviary Birds of India,* " to this lovely bird (from a fancier's point of view), and that is its very savage temper in some cases. In the wild state Mr. E. C. Stuart Baker has seen two of these birds fight to death, and another couple defy law and order by hustling a king-crow, of all birds. And in confinement it is difficult to get two to live together ; while some specimens are perfectly impossible companions for other small birds, savagely driving them about and not allowing them to feed. Many individuals, however, are quite peaceable with other birds, and a true pair will live together in harmony."

There is nothing remarkable in the nest of the *Chloropsis;* it is a shallow cup, devoid of lining, placed fairly high up in a tree. July and August are the months in which to look for nests. Two eggs usually form the complete clutch. It would thus seem that green bulbuls have not a great many enemies to fear. Nevertheless they fuss as much over their eggs as some elderly ladies of my acquaintance do over their baggage when travelling. Birds and people who worry themselves unduly over their belongings seem to lose these more often than do those folk who behave more philosophically. Take the case of the common bulbuls. These certainly lose more broods than they succeed in rearing, yet the ado they make when a harmless creature approaches their nest puts one forcibly in mind of the behaviour of the captain of a Russian gunboat when an innocent vessel happens to enter the zone of sea in the centre of which the Czar's yacht floats.

XXII

CORMORANTS

CORMORANTS, like Englishmen, have spread themselves all over the earth. Save for a few out-of-the-way islands, there is no country in the world that cannot boast of at least one species of cormorant. Cormorants, then, are an exceedingly successful and flourishing family. It must be very annoying for those worthy professors and museum naturalists who are always lecturing to us about the all-importance of protective colouration that the most flourishing families of birds—the crows and the cormorants—are as conspicuous as it is possible for a thing in feathers to be.

Mr. Seton Thompson well says that every animal has some strong point, or it could not exist ; and some weak point, or the other animals could not exist. Cormorants have several strong points, and that is why they flourish like the green bay tree, notwith standing their conspicuous plumage. They are as hardy as the Scotchman, as voracious as the ostrich, as tenacious of life as a cat, to say nothing of being piscatorial experts, powerful fliers, and champion divers.

The cormorant family furnishes a very good example of the manner in which new species arise quite independently of natural selection. Notwithstanding their world-wide distribution, all cormorants belong to one genus, which is divided up into thirty-seven species. Of these no fewer than fifteen occur in New Zealand—a country not characterised by a large avifauna.

One species—the large cormorant (*Phalacocorax carbo*)—flourishes in almost every imaginable kind of climate and among all sorts and conditions of birds and beasts. Yet in New Zealand, in a country where the conditions of existence vary but little, cormorants have split up into fifteen species. It is therefore as clear as anything can be in nature that we must look to some cause other than natural selection for an explanation of the multiplicity of species of cormorant in New Zealand. It seems to me that the solution of this puzzle lies in the fact that the conditions of life are comparatively easy in New Zealand. Consequently a well-equipped bird like a cormorant is allowed a certain amount of latitude as to its form and colouring. In places where the struggle for existence is very severe, where organisms have their work cut out to maintain themselves, the chances are that every unfavourable variation will be wiped out by natural selection ; but if the struggle is not particularly severe, or if a species has something in hand, it can afford to dispense with part of its advantage and still survive. Thus it is that in New Zealand we see a number of different species of cormorant living side by side. De Vries likens natural selection to a sieve through which all organisms are sifted, and through the meshes of which only those of a certain description are able to pass. Bateson compares it to a public examination to which every organism must submit itself. Those animals that fail to get through are killed. The standard of the examination may vary in various parts of the world.

So much for cormorants in general and the puzzle they present to evolutionists. Let us now consider for a little while our Indian cormorants. For once India is at a disadvantage as compared with New Zealand. There are but three species found in this country—the great, the lesser, and the little cormorant. The last—*Phalacocorax javanicus*—is the most commonly seen of them all. It is to be found in the various backwaters round about Madras, being especially abundant in the vicinity of Pulicat. At the place where the canal runs into the lake there are a number of stakes driven into the canal bed; these project above the level of the water, and on every one of them a little cormorant is to be seen. Cormorants in such a position always put me in mind of the pillar saints of ancient times. Although very active in the water, cormorants become statuesque in their stillness when they leave it.

The lesser cormorant (*Phalacocorax fuscicollis*) breeds in nests in the trees on the islets which stud the Red-hills Tank near Madras, also on the tank at Vaden Tangal, near Chingleput. The third species of cormorant found in India is the great cormorant (*Phalacocorax carbo*). This is the one which is world-wide in its distribution. It is a large bird, being over 2 ft. 6 in.

in length. It is said to be capable of swallowing at one gulp a fish fourteen inches long. It is less gregarious in its habits than the other cormorants, but it breeds and roosts in colonies. Captain H. Terry states that this species' nests are to be met with on a tank near Bellary. The great cormorant possesses fourteen tail feathers, while all other cormorants have to put up with twelve. Why the big fellow should be the happy possessor of two extra caudal feathers is a puzzle which no one has attempted to solve.

It is not very easy to distinguish the three species of cormorant from one another. The great cormorant has a conspicuous white bar on each side of the head. This and his larger size serve to separate him from the two smaller forms. It is usually possible to distinguish the other two by the fact that the little cormorant has more white on the throat than his somewhat larger cousin. But, when all is said and done, it is not of great importance to distinguish the various species. All cormorants have almost exactly the same habits. The nests are all mere platforms of sticks. They are all expert fishermen, and seem equally at home on fresh or salt water. They can swim either on or under water and move at a considerable pace, covering nearly 150 yards in a minute. The young are said to feed themselves by inserting their heads into the gullet of the parent and pulling out the half-digested fish. Cormorants are readily tamed and are employed in China to fish for their masters, a rubber ring being inserted round the lower part of the neck in order to prevent the fish from going too far. In bygone days, fishing by means of cormorants was considered good sport, and the royal household used to have its Master of the Cormorants.

. Cormorants' eggs are of a very pale green colour, and their nests smell of bad fish, for the owners care nothing about sanitation. Young cormorants are not nearly so black as their parents, and do not attain adult plumage till they are four years old.

XXIII

A MELODIOUS DRONGO

OUR friend the king crow (*Dicrurus ater*) is so abundant throughout India, and possesses to so great a degree the faculty of arresting the attention, that we are apt to overlook his less numerous relatives. In Ceylon it is otherwise. *Dicrurus ater* occurs in that fair isle, but only in certain parts thereof, and is not so abundant as his cousin, the white-vented drongo (*Dicrurus leuco-pygialis*). The former has, therefore, to play second fiddle in Ceylon, where he is usually known to Europeans as the black fly-catcher. The white-vented drongo is their king crow—the bird that lords over the *corvi*.

The drongos constitute a well-defined family. When you know one member you can scarcely fail to recognise the others. They fall into two great classes, the fancy and the plain, the dandies and those that dress quietly. The bhimraj, or larger racket-tailed drongo (*Dissemurus paradiseus*), is the most perfect example of the fancy or ornamental class. His head is set off by a crest, but his speciality is the pair of outer tail feathers, which attain a length of nearly two feet.

Of the less ornamental drongos, the king crow is the best-known example. This bird is found in all parts of India, and occurs in Ceylon. Almost as widely distributed, but far less abundant, is the white-bellied drongo. This species may be met with in all parts of India save the Punjab. In the Western Province of Ceylon it is replaced by a drongo having less white in the plumage.

It is a moot question whether this last is to be looked upon as a race or a distinct species. Legge writes : " No bird in Ceylon is so puzzling as the present, and there is none to which I have given so much attention with a view to arriving at a satisfactory determination as to whether there are two species in the island or only one. I cannot come to any other conclusion than that there is but one, the opposing types of which are certainly somewhat distinct from one another, but which grade into each other in such a manner as to forbid their being rightly considered as distinct species ; and I will leave it to others, who like to take the matter up for investigation, to prove whether my conclusions are erroneous or not." Oates has since constituted the birds which have less white on the lower parts a distinct species, which he calls the white-vented drongo (*Dicrurus leucopygialis*). He admits that the amount of white on this form and on the white-bellied species (*Dicrurus cærulescens*) is variable, and that a bird is occasionally met with which might, as regards this character, be assigned indifferently to one or the other species, but, says he, the colour of the throat and breast will, in these cases, be a safe guide in identification. The parts in question are grey in the white-bellied species and dark brown in the white-vented form. It seems to me that a slight difference in the colouring of the feathers of the throat is not a very safe foundation on which to establish a new species. However, this piece of species-splitting need . not worry the Anglo-Indian, for the white-vented form is found only in Ceylon. All drongos with white underparts that occur in India are *Dicrurus cærulescens*. This bird is not common in Madras ; I observed it but twice during eighteen months' residence in that city. It is in shape exactly like the common king crow, and possesses the characteristic forked tail, but it is a smaller bird, being nine and a half inches in length, and therefore shorter by fully three inches than the black drongo. Its upper plumage is deep indigo ; the throat and breast are grey ; all the remainder of the lower plumage is white. Its habits are very much like those of the king crow, but it is less addicted to the open country, seeming to prefer well-wooded localities. I have never seen the *Dhouli*, or white-bellied drongo, perched on anything but a branch of a tree. It almost always catches its insect prey upon the wing, after the manner of a fly-catcher. Jerdon, however, states that he once saw it descend to the ground for an insect.

As a singer it is far superior to the king crow. In addition to the harsh notes of that species it produces many melodious sounds. Tickell describes its song as " a wild, mellow whistle pleasingly modulated." It was the voice of the bird that first attracted my notice. Some eight years ago, when camping in the Fyzabad

District, I heard a very pleasing but unknown song. Tracking this to the mango tope whence it issued, I discovered that the author was a white-bellied king crow. Last winter a member of this species favoured me with a fine histrionic performance. I was sitting outside my tent one afternoon, when I heard above me a harsh note that was not quite like that of the king crow. Looking up, I observed, perched on a bare branch at the summit of the tree, a white-bellied drongo. Then, as if for my especial benefit, he began to imitate the call of the shikra ; he followed this up by a very fair reproduction of some of the cries of a tree-pie. Having accomplished this, he made, first his bow, then his exit. I was much interested in the performance, since an allied species, the bhimraj, is not only one of the best songsters in the East, but a mimic second only to the wonderful mocking-bird of South America.

The white-bellied drongo is so rare in the peninsula of India that not one of our ornithologists has given us anything like a full account of its habits, and no one appears to have discovered the nest in India. Fortunately, it is very common in Ceylon, so that Legge has been able to give some interesting details regarding its habits. We must bear in mind that Legge includes both the white-bellied varieties under one species. If we divide them into two, the question arises to which do his various observations apply ? The reply is to either or both, for Legge was not able to detect any differences between them, except that perhaps the white-vented variety has a more powerful voice. He

K

writes : " It is occasionally, when there is abundance of food about, a sociable species, as many as three or four collecting on one tree, and carrying on a vigorous warfare against the surrounding insect world." Like the king crow, it is an early riser and a late rooster. It is a great chaser of crows, and of any creature that dares to intrude into the tree in which its nest is placed. Needless to say that it detests owls. Says Legge : " The white-bellied king crow never fails to collect all the small birds in the vicinity whenever it discovers one of these nocturnal offenders, chasing it through the wood until it escapes into some thicket which baffles the pursuit of its persecutors." But why does he call owls " nocturnal offenders " ? Wherein lies their offence ? So far as I can see, the only crime that owls commit is in being owls. The creatures they prey upon have reason for disliking them. But owls do not attack ornithologists. Why, then, should these gentry call them hard names ?

The nesting habits of both the white-bellied and the white-vented drongos are very similar to those of the common king crow. Legge describes the nest as a shallow cup, almost invariably built at the horizontal fork of the branch of a large tree at a considerable height from the ground, some-times as much as forty feet. The eggs seem to vary as greatly in appearance as do those of the common king crow.

Since the white-bellied drongo appears to be quite as pugnacious as its black cousin, and to have almost identical habits, it is strange that it should be so uncommon in India. As we have seen, its distribution is wide, so that it seems able to adapt itself to various kinds of climate. Nevertheless, it is common nowhere in India. What is the cause that keeps down its numbers? Naturalists are wont to talk airily about natural selection causing a species to be numerous or the reverse, but unless they can show precisely how natural selection acts they explain nothing. Those who write books on natural history convey the impression that it is the birds and beasts of prey that keep down the numbers of the smaller fry. As a matter of fact, predaceous creatures seem to exercise but little influence on the numbers of their quarry. There are hidden causes at work of which we know almost nothing. Damp and small parasites are probably far more powerful checks on multiplication than predaceous creatures. It would seem that there is something in the constitution of the white-bellied drongo which enables it to outnumber the king crows proper in Ceylon, but which prevents it from becoming abundant in the peninsula of India. What this something is we have yet to discover. We really know very little of the nature of that mysterious force with which naturalists love to conjure, and which Darwin named Natural Selection. We write it with a capital N and a capital S, and then imagine that we have explained everything.

> " Twinkle, twinkle, little star,
> Now we all know what you are."

XXIV

THE INDIAN PITTA

SOME Indian birds are adepts at self-advertisement. To use an expressive vulgarism, they continually " hit you in the eye "; they obtrude themselves upon you in season and out of season. Others are so retiring that you may live among them for years without observing them. To this class, to the class that hide their light under a bushel, the beautiful Indian pitta (*Pitta brachyura*) belongs. There is at least one favoured compound in Madras where a pitta, or possibly a pair of them, spends the cool-weather season. Pittas proclaim their presence by uttering at dawn their cheery notes, which have been described as an attempt to whistle, in a moderately high key, the words " quite clear." If, on hearing this call, you are sufficiently energetic to go out of doors, you will probably see on the ground a bluish bird, about the size of a quail, but before you have had time to examine it properly it will have taken to its wings and disappeared into the hedge. Those who are not so fortunate as to have pittas on the premises may be tolerably certain of seeing a specimen by visiting the well-wooded plot of land bordered on the west by the canal and on the south by the Adyar River.

This bird is about seven inches in length. Thus it does not measure much more than a sparrow, but it is nevertheless considerably larger, for the tail is very short, being barely one inch and a half in length. The crown is yellow, tinged with orange, and divided in the middle by a broad black band running from the beak to the nape of the neck, where it meets a broader black band that passes below the eye. The eyebrow is white. The back and shoulders are dull bluish green. The upper tail-coverts are pale blue. There is also a patch of this colour on the wing. The wings and tail are black, tipped with blue. During flight the pinions display a white bar. The chin and throat are white. The breast is of the same yellow hue as the head. There is a large crimson patch under the tail. Captain Fayrer's photograph in *Bombay Ducks* conveys very well the shape of the bird, but, of course, does not reproduce the most marked feature of the pitta—its colouring. Indians in some localities call it the *naurang* —the nine-colours. The bird may truly be said to be arrayed in a coat of many colours. Unfortunately, such a garment is apt to lead to trouble. Even as the coat of many colours brought tribulation upon Joseph of old, so does the much-coveted, multi-hued plumage of the pitta frequently bring death to its possessor.

Apart from the colouring, it is impossible to confound the pitta with any other bird. Its long legs and its apology for a tail recall the sandpiper, but there is nothing else snippet-like about it. The classification of the bird has puzzled many a wise head. It has been variously called the Madras jay, the Bengal quail, the short-tailed pye, the ant-thrush, the painted thrush, and the ground thrush. But it is not a jay, neither is it a quail, nor a thrush, nor a tailless pye. It is a bird made on a special model. It belongs to a peculiar family, to a branch of the great order of perching birds, which differs from all the other clans in some important anatomical details. Into these we will not go, for they belong to morphology, the science which concerns itself chiefly with the dry bones of zoology, with the lifeless aspect of the science of life.

The Indian pitta is a bird which likes warmth, but not heat, so that it refuses to live in the Punjab, where the climate is one of extremes—a spell of cold, then a headlong rush into a period of intense heat, followed by an equally sudden return to a low temperature. The pitta seems to occur in all parts of Eastern, Central, and Southern India, undergoing local migration to the south in the autumn and back again in the spring. In places where the climate is never very hot or very cold, as, for example, Madras and the hills in Ceylon, some individuals seem to remain throughout the year. I have seen pittas in Madras at all seasons, and I know of no better testimonial to the excellence of the climate of that city. Jerdon writes of the pitta : " In the Carnatic it chiefly occurs in the beginning of the hot weather, when the land-winds first begin to blow with violence from the west ; and the birds in many instances appear to have been blown, by the strong wind, from the Eastern Ghauts, for, being birds of feeble flight, they are unable to contend against the strength of the wind. At this time they take refuge in huts, outhouses, or any building that will afford them shelter. The first bird of this kind that I saw had taken refuge in the General Hospital at Madras ; and subsequently, at Nellore, I obtained many alive under the same circumstances." Other observers have had similar experiences. Bligh, for instance, states that in Ceylon pittas are frequently caught in bungalows on coffee estates on cold and stormy days.

It is strange that so retiring a bird as the pitta should find its way with such frequency into inhabited houses. Jerdon's explanation is its " feeble flight," but I doubt whether he is correct in calling the pitta a bird of weak flight ; it can travel very fast, for short distances at any rate. It seems to me that the pitta dislikes cold and wind, and therefore naturally seeks any shelter that presents itself. Not being a garden bird, it is unaware that the bungalow, which offers such tempting cover, is the abode of human beings. Possibly another reason why the pitta so frequently enters bungalows is to avoid the crows. Dr. Henderson tells me that he was playing tennis some years ago at a friend's house in Madras when he saw a bird being chased by a mob of crows. The fugitive took refuge in the drawing-room of the house, where Dr. Henderson caught it, and found that it was an uninjured but very much frightened pitta. Mr. D. G. Hatchell informs me that he once picked up in his verandah a dead pitta that had probably been killed by crows. The *corvi* are out-and-out Tories. They strongly resent all innovation *qua* innovation. Any addition to the local fauna is exceedingly distasteful to them. They object to the foreigner quite as strongly as do (perhaps I should say " did ") the Chinese. It is for this reason that they mob every strange bird that shows its face. Now, they seldom come across either the creatures of the night or the denizens of the thick undergrowth ; consequently, when such venture forth into the light of day the crows forthwith attack them.

The pitta feeds chiefly on beetles, termites, ants, and other creeping things, which it seeks out among fallen leaves, after the manner of the " seven sisters." The pitta is quick on its feet, and is able to hop and run with equal ease. It thrives in captivity. It is an excellent pet, provided it be not kept with smaller birds. It regards these as so much fresh meat especially provided for it.

The nest of the pitta is described as a globular structure fully nine inches in horizontal diameter and six inches high, with a circular aperture on one side. Twigs, roots, and dried leaves are the building materials utilised. The eggs are exceedingly beautiful. " The ground colour," writes Hume, " is China white, sometimes faintly tinged with pink, sometimes creamy ; and the eggs are speckled and spotted with deep maroon, dark purple, and brownish purple as primary markings, and pale inky purple as secondary ones. Occasionally, instead of spots, the markings take the form of fine hair-like lines."

XXV

THE INDIAN WHITE-EYE

THE Indian white-eye (*Zosterops palpebrosa*) is a bird which should be familiar to everyone who has visited the Nilgiris. To wander far in a hillside wood without meeting a flock of these diminutive creatures is impossible. Sooner or later a number of monosyllabic notes will be heard, each a faint, plaintive cheep. On going to the tree from which these notes appear to emanate a rustle will be observed here and there in the foliage. Closer inspection will reveal a number of tiny birds flitting about among the leaves. These are white-eyes—the most sociable of birds. Except when nesting, they invariably go about in companies of not less than twenty or thirty. Each individual is as restless as a wren, so that some patience must be exercised by the observer if he wish to obtain a good view of any member of the flock. But by standing perfectly motionless for a time under the tree in which the birds are feeding he who is watching will, ere long, be able to make out that the white-eye is a tiny creature, not much more than half the size of a sparrow. The upper parts are yellowish green, the chin, throat, and feathers under the tail are bright yellow, and the remainder of the lower surface of whitish hue. The most marked feature of the *Zosterops* is a conspicuous ring of white feathers round the eye, which causes the bird to look as though it were wearing white spectacles. From this circle the species derives its popular names, the white-eye or spectacle bird. Thanks to the conspicuous eye-ring, it is impossible to mistake the bird.

All feathered creatures that go about in flocks and haunt thick foliage emit unceasingly a call note, by means of which the members of the flock keep in touch with one another. This ceaseless cheeping note is probably uttered unconsciously. Each individual listens, without knowing that it is doing so, for the calls of its fellows; so long as it hears these it is happy. When the main volume of the sound grows faint the individual white-eye knows that his companions are moving away from him; he accordingly flies in the direction from which their calls are coming, giving vent, as he goes, to a louder cheep than usual. Whenever a white-eye flies from one tree to another it utters this more powerful call and thereby informs its fellows that it is moving forward. This louder cry stimulates the others to follow the bird that has taken the lead. All the time they are thus flitting about the white-eyes are busy picking tiny insects off the leaves. I have never observed them eating anything but insects. Legge, however, asserts that their diet is for the most part frugivorous, in consequence of which the birds are, according to him, very destructive to gardens, picking off the buds of fruit trees, as well as attacking

the fruit itself. He further declares that he has
known caged individuals in England feed with avidity
on dried figs. Hutton also states that white-eyes
feed greedily upon the small black berries of a species
of *Rhamnus*, common in the Himalayas. Notwithstanding the authorities cited, it is my belief that these
little birds are almost exclusively insectivorous. They
perform a useful work in devouring numbers of ob-
noxious insects, which they extract from flowers. In
so doing their heads sometimes become powdered
with pollen, so that the white-eyes probably, like bees
and moths, render service to plants by carrying pollen
from one flower to another.

The search for food does not occupy the whole day.
Except at the nesting season, the work of birds is light.
In the early morning the white-eyes feed industriously ;
so that by noon they have satisfied their hunger. They
then flit and hop and fly about purely for pleasure.
White-eyes, like all small birds, literally bubble over
with energy. They are as restless as children. Once
when walking through the Lawrence Gardens at
Lahore in the days when they had not yet fallen into
the clutches of that enemy of beautiful scenery, the
landscape gardener, I came across a company of these
charming little birds disporting themselves amid some
low bushes near a plantation of loquat trees. First
one little bird, then another, then a third, a fourth, a
fifth, etc., dropped to the ground, only to return at once
to the bush whence they came. A whole flock appeared to be taking part in this pastime. There were
two continuous streams, one of descending and the

other of ascending white-eyes. These might have been little fluffy, golden balls with which some unseen person was playing.

When the heat of the day is at its zenith, white-eyes, like most birds in India, enjoy a *siesta*. At this hour little gatherings of them may be seen, each bird huddled against its neighbours on some bough of a leafy tree.

At the nesting season the white-eye sings most sweetly. The ordinary cheeping note then becomes glorified into something resembling the lay of the canary ; less powerful, but equally pleasing to the ear.

The nest of the white-eye is a neat little cup, or, as Mr. A. Anderson describes it, a hollow hemisphere. It is a miniature of the oriole's nursery. It is large for the size of the bird, being usually over two inches in diameter. Some nests are fully two inches deep, while others are quite shallow. It is composed of fine fibres (i.e. grass stems, slender roots, moss, and seed down) and cotton, bound together by cobweb, which is the cement most commonly used by bird masons. The nursery is invariably provided with a lining. In one nest that I found, this lining consisted of human hair. Other lining materials are silky down, hair-like moss and fern-roots, and grass fibres so fine that the horsehairs which are sometimes utilised look quite coarse beside them. The most wonderful thing about the nest of this pretty little bird is the manner in which it is attached to its supports. I have called it a minia-ture of the oriole's nursery, because it is usually sus-pended from two or more branches by cotton fibres. I once came upon a nest which was attached to but

one slender branch, and to the tip of this. The end
was worked into the structure of the nest so that the
whole looked like a ladle with a very thin handle. It
seemed incredible that so slight a branch could support
the nest and its contents.

I have not been fortunate enough to watch the
white-eye building its nest. Mr. A. Anderson states
that the pair—for both the cock and the hen take
part in nest construction—" set to work with cobwebs,
and having first tied together two or three leafy twigs
to which they intend to attach their nest, they then use
the fine fibre of the *sunn* (*Crotalaria juncea*), with which
material they complete the outer fabric of their very
beautiful and compact nest. As the work progresses,
more cobwebs and fibre of a silky kind are applied
externally, and at times the nest, when tossed about
by the wind (sometimes at a considerable elevation),
would be mistaken by a casual observer for an acci-
dental collection of cobwebs. The inside of the nest
is well felted with the down of the *madar* plant, and
then it is finally lined with fine hair and grass stems
of the softest kind." The nest is usually situated
within three or four feet of the ground, but is sometimes
placed at much higher elevations.

In South India, the time to look for white-eyes'
nests is from January to March. In the north, the
majority of nests are found between April and June.

The eggs are a beautiful pale blue. Most commonly
only two seem to be laid. There are, however, many
cases on record of three and a few of four eggs. This
is an unusually small clutch. Nevertheless it is un-

likely that a pair of white-eyes bring up more than two broods in the year. These facts, when taken in conjunction with the wide distribution of the species, indicate that the white-eye meets with exceptional success in rearing its young. The nest is usually well concealed in the depths of a leafy bush. Squirrels and lizards must find the suspended nursery difficult of access. In addition to this we must bear in mind that white-eyes are plucky little creatures. Mr. Ball describes how he saw one of them attacking a rose-finch, a vastly more powerful bird, and drive it away from the flowers of the *mohwa*, which form a favourite hunting-ground of the white-eye.

As I have repeatedly stated, pugnacity is a more valuable asset than protective colouration in the struggle for existence.

Lastly, the white-eye appears able to thrive under greatly varying conditions of climate.

These advantages possessed by white-eyes, I think, explain why the clutch of eggs is so small.

White-eyes make excellent pets. They will live amicably along with amadavats in a cage. Finn is my authority for saying that soft fruit, bread and milk, and small insects are all the food required by white-eyes, and they are so easy to keep that many specimens are sent to Europe.

XXVI

GOOSEY, GOOSEY GANDER

THE goose, like certain ladies who let lodgings, has seen better times. It is a bird that has come down in the world. For some reason which I have never been able to discover, it is nowadays the object of popular ridicule. It is commonly set forth as the emblem of foolishness. Invidious comparisons are proverbially drawn between it and its more handsome cousin, the swan. The modern bards vie with one another in blackening its character. As Phil Robinson says, " It does not matter who the poet is—he may be anyone between a Herbert and a Butler—the goose is a garrulous fool, *et præterea nihil.*" Well may the bird cry *O tempora ! O mores !* It has indeed fallen upon evil days. Things were not ever thus. Time was when men held the goose in high esteem. Livy was loud in his praises of the bird. Pliny was an ardent admirer thereof. The Romans used to hold a festival in honour of the feathered saviour of the Capitol. The degradation of the goose is, I fear, a matter of looks. Its best friend can scarcely call it handsome. It is built for natation rather than perambulation ;

143

nevertheless it spends much time out of water and feeds chiefly on *terra firma*. It is probably a bird that is undergoing evolution, a bird that is changing its habits. It has taken to a more or less terrestrial existence, but has not yet lost what I may perhaps call the aquatic waddle. While walking it looks as though it might lose its balance at any moment. As a matter of fact, the goose is no mean pedestrian, and is capable of performing considerable journeys on foot. When pressed, it can show a fine turn of speed. This I have had some opportunity of observing in the Zoological Gardens at Lahore. A crane (*Grus antigone*) is confined in the water-birds' enclosure along with the ducks, geese, pelicans, etc. Now, cranes are the most frolicsome and playful of birds. In no other fowl is the sense of humour more highly developed. The crane in question continually indulges in " cake walks," and cuts other mad capers. Sometimes it is seized with the impulse to " clear the decks," that is to say, the banks of the ornamental pond. The operation is conducted as follows : The crane opens out its wings, takes two wild jumps into the air, then rushes at the nearest duck or goose, with wings expanded, looking as though it were going through one of the figures of the serpentine dance. The frightened duck flees before the crane ; the latter keeps up the chase until the duck takes refuge in the water. Having succeeded in its object, the crane trumpets loudly and performs a dance which a Red Indian on the war-path could scarcely hope to emulate. It next turns its attention to some other inoffensive duck or goose. It is while being thus chased that pinioned geese show a fine turn of speed. Fly they cannot, so they sprint with expanded wings.

The goose is a great favourite of mine. The more one sees of the bird the more one likes it and appreciates its good qualities. It is a creature of character. It rapidly forms attachments, and will sometimes follow about, like a dog, the person to whom it has taken a fancy. A curious instance of this was recorded many years ago by *The Yorkshire Gazette*. A gander belonging to a farmer developed a liking for an old gentleman. The bird used to go every morning from the farmyard to the house of the said elderly gentleman and awake him by its cries. It would then accompany him the whole day in his walks and strut behind him in the most frequented streets, unmindful of the screams of the urchins by whom the strange pair were often followed. When the old gentleman sat down to rest the gander used to squat at his feet. When they were approaching a seat on which the old man was accustomed to sit the gander used to run on ahead and signify by cackling and flapping of wings that the resting-place was reached. When anyone annoyed the old gentleman the gander would express its displeasure by its cries and sometimes by biting. When its friend went into an inn to take a glass of ale, the bird used to follow him inside if permitted ; if not allowed to do so, it would wait outside for him.

One should not of course accept as gospel truth everything one reads in a newspaper. It is necessary to discriminate. Thus, when a well-known weekly journal

L

produces a picture of the ladies of a Sultan's harem dancing unveiled before a distinguished company of gentlemen, one begins to wonder whether truth really is stranger than fiction. However, I see no reason to doubt the substantial accuracy of the story of the Yorkshire gander. The goose is an exceptionally intelligent bird and is very easily tamed. I once made friends with a goose in the Zoological Gardens at Lahore. It was a white, bazaar-bred bird. Whenever it saw me it used to walk up to the fence and emit a low note of welcome. I was able to distinguish that particular bird from the other geese by the fact that a piece had been broken off its upper mandible.

I am glad to notice that Mr. W. H. Hudson, one of the leading British ornithologists, has a high opinion of the goose. In his *Birds and Man* he gives a delightful account of the home-coming of a flock of tame geese led by a gander. He writes : " Arrived at the wooden gate of the garden in front of the cottage, the leading bird drew up square before it, and with repeated loud screams demanded admittance. Pretty soon in response to the summons, a man came out of the cottage, walked briskly down the garden-path and opened the gate, but only wide enough to put his right leg through ; then placing his foot and knee against the leading bird he thrust him roughly back ; as he did so three young geese pressed forward and were allowed to pass in ; then the gate was slammed in the face of the gander and the rest of his followers, and the man went back to the cottage. The gander's indignation was fine to see, though he had probably experienced the same rude treatment on many previous occasions. Drawing up before the gate again, he called more loudly than before ; then deliberately lifted a leg, and placing his broad webbed foot like an open hand against the gate, actually tried to push it open. His strength was not sufficient, but he continued to push and call until the man returned to open the gate and let the birds go in."

If only for his sturdy independence and his insistence on his rights the gander is a bird whose character is worthy of study. He is courageous too ; so is his wife. She will stand up fearlessly to a boy, a kite, or even a fox, when her brood is threatened. Last year in the Lahore Zoological Gardens a goose hatched a number of goslings. The kites regarded these as fair game, and, in spite of the efforts of the mother, carried off several of the young birds. Thereupon four ganders took counsel and constituted themselves a bodyguard for the goose and chicks, one or more of them being always on duty. In spite of this a kite managed to secure another gosling. The mother and her remaining five chicks were then placed in a cage ; notwithstanding this, the ganders still main tained their guard and cried loudly whenever a human being approached the cage containing the brood.

The goose, like the swan, uses its wing as a weapon. When it attacks it stretches its neck and head low along the ground and hisses ; it then dashes at its adversary, seizes him with beak and claws, and lays on to him right well with its powerful wings.

Here endeth the account of " Goosey, Goosey Gander." I must apologise to the geese in their natural state for having completely ignored them. We will make amends by indulging in a wild-goose chase at an early date !

XXVII

GEESE IN INDIA

SEVEN or eight species of goose have been recorded as winter visitors to India. With two exceptions they honour us with their presence only on rare occasions, and do not really form part and parcel of our Indian avifauna. The exceptions are the grey lag goose and the barred-headed goose, which visit India every winter in their millions. It is these that form the subject of this essay. It is difficult for the dweller in the south to realise how abundant geese are in Northern India throughout the cold weather. Flocks of them fly overhead so frequently that they scarcely attract notice. Each flight looks like a great trembling, quivering V, floating in the air, a V of which the angle is wide and one limb frequently longer than the other. During flight geese are distinguishable from cranes and storks by this V-shaped formation, and by the fact that they never sail on expanded wings ; they progress by means of a steady, regular motion of the pinions, and are able to cover long distances in short time. Geese on the wing are distinguishable from the smaller species of duck by their larger size, and from Brahminy ducks (*Casarca rutila*) by their lighter colour. Moreover, the curious note of this last species is very different from the cackle of geese. Brahminy ducks go about in couples ; geese fly in flocks.

Like most birds which breed in the far north, geese are largely nocturnal ; their cries as they fly overhead are among the commonest of the sounds which break the stillness of the winter night in Upper India. They feed mainly in the hours of darkness, and do a certain amount of damage to the young wheat ; nor do they leave their feeding ground until the sun is high in the heavens, when they repair to a river bank or shallow lake, where they love to bask in the sun, all with the head tucked under the wing, save one or two who do duty as sentinels.

The grey lag goose of India is, I believe, identical with the wild goose of England. This is a belief not shared by everyone. For over a century this species has been the plaything of the systematist. Linnæus classed ducks and geese as one genus—*Anas*. This goose he called *Anas anser*, the goose-duck. But it was soon recognised that ducks and geese are not sufficiently nearly related to form a common genus ; hence, the geese were formed into the genus *Anser,* and the grey lag goose was then called *Anser cinereus,* the ashy-coloured goose, a not inappropriate name, although the bird is brown rather than grey. But the name was not allowed to stand. For some reason or other it was changed to *Anser ferus.* Then it was altered to *Anser anser*—the goosey goose, presumably meaning the goose *par excellence.* Then Salvadori discovered, or thought that he discovered, that the grey lag geese of the East are not quite like those of the West ; he therefore made two species of the bird, calling the Indian variety *Anser rubirostris*, the red-billed goose. Alphéraky denies the alleged difference. The result is that the bird has some half-dozen names, each of which has its votaries. It is this kind of thing which deprives classical nomenclature of all its utility. Until ornithologists learn to grasp the simple fact that the external appearance of every living creature is the result of two sets of forces, internal or hereditary, and external or the influence of environment, they will always be in difficulty over species. Englishmen who dwell in sunny climates get browned by the sun, yet no one dreams of making a separate species of sun-burned Englishmen. Why, then, do so in the case of birds whose external appearance is slightly altered by their environment ?

Even as scientific men have toyed with the Latin name of the bird, so have compositors played with its English name. Nine out of ten of them flatly decline to call it the grey lag goose ; they persist in setting it down as the grey *leg* goose. If the bird's legs were grey this would not matter. Unfortunately they are not. In extenuation of the conduct of the compositor there is the fact that etymologists are unable to agree as to the meaning and derivation of the word lag.

The other common species of goose is the barred-headed goose (*Anser indicus*). This is not found in Europe. It is a grey bird, more so than the grey lag goose, with two black bands across the back of the head.

The upper one runs from eye to eye, the lower is parallel to, but shorter than, the upper bar. The back of the neck is black and the sides white. There is some black in the wings. The bill and feet are yellow. Both these species of goose are a little smaller than the domestic bird.

Geese are very wary creatures, and possess plenty of intelligence. They all seem to know intuitively the range of a gun, and as they object to being peppered with No. 2, or any other kind of shot, it is necessary for the sportsman to have recourse to guile if he would make a bag. It is this which makes shooting them such good sport. Every bird obtained has to be worked for. By rising very early in the morning the gunner may sometimes get a shot at them while they are feeding. They seem to be less wary then than later in the day. Sometimes, when riding at sunrise, I have suddenly found myself within forty yards of a flock of geese feeding in a field.

They usually indulge in their midday siesta in an open place, and invariably post sentinels. For this reason they do not give one much opportunity of observing them. They cannot, or pretend they cannot, distinguish between the naturalist and the sportsman. In this, perhaps, they are wise. Their intelligence has, I think, been exaggerated. Last winter, when punting down the Jumna, I noticed a flock of geese resting on the moist sand at the water's edge. Behind them was a semi-circular sandbank, some fifteen feet in height. This bank sheltered the geese from the wind. Birds, like ladies, object to having their feathers ruffled. It occurred to me that owing to the sandbank one could approach quite near to the flock unobserved. Knowing that geese are creatures of habit, I counted on the flock being at the identical spot next day. Consequently, I returned the following morning and approached on all fours from the sandbank side, and was rewarded by securing a barred-headed goose. I repeated the operation on the following day, and again bagged a goose. The third day I was unable to visit the place, so sent a friend, who was only prevented from slaying a goose by the fact that two Brahminy ducks in midstream saw him approaching and gave the alarm. We left the camp the next day. I do not, therefore, know whether the geese continued to frequent that danger-fraught sandbank. The fact that they allowed themselves to be caught napping thrice shows that they have not quite so much intelligence as some people credit them with. For all that, the goose is no fool.

XXVIII

A SWADESHI BIRD

I COMMEND the common peafowl (*Pavo cristatus*) to the Indian patriot, for it is a true *Swadeshi* bird. It is made in India and nowhere else. The beastly foreigner does, indeed, produce a cheap imitation in the shape of *Pavo muticus*—the Javan peafowl; but with this the patriotic Indian bird will have nothing to do. The two species are very like in appearance, the most noticeable difference being in the shape of the crest; that of the Indian species is like an expanded fan, while the cranial ornament of the Javan species resembles a closed fan. Notwithstanding their similarity they do not interbreed when brought together. This, I am aware, was not Jerdon's view. He stated that hybrids between the two species were not rare in aviaries. In this particular instance Jerdon, *mirabile dictu*, seems to have been wrong; he probably mistook the japanned variety of the common bird for a hybrid. My experience tends to show that the two species will not interbreed. Caste feeling evidently runs high.

Peafowl are distributed all over India; they occur in most localities suited to their habits, that is to say where there is plenty of cover, good crops, and abundance of water. They are very plentiful in the Himalayan *terai*, where they are a source of annoyance to the sportsman. You are sitting in your *machan*, listening to the approaching line of beaters. Presently there is a rustle among the fallen leaves ; a creature is making its way through the undergrowth. You listen intently, and perceive with satisfaction that the moving object is coming towards your *machan*. You are now all attention, and grasp your rifle in such a manner that it can, in an instant, be brought to your shoulder. Then, to your disgust, a peacock emerges with a good-morning-have-you-used-Pear's-soap air. When, after about half a dozen of these false alarms, a bear appears, you are, as likely as not, unprepared for him.

In many parts of Northern India, notably in those districts through which the Jumna and Ganges run, peafowl are accounted sacred by the Hindu population. If you shoot one in such a locality the villagers have a disagreeable way of turning out *en masse*, armed with *lathis*. The reverence for the peacock is curiously local. In one village the people will invite you to shoot the birds on account of the damage they do to the crops ; while the inhabitants of a village at a distance of less than a hundred miles will send a wire to the Lieutenant-Governor if you so much as point a gun at the sacred fowl. I once camped in a district where peafowl were exceptionally numerous, and on this account I concluded that they were venerated by the populace. But, sacred or not, I hold that there is nothing to equal a young peafowl as a table bird, so I used to mark down the trees in which the pea-chicks roosted, and return to the spot with a gun, after the shades of night had fallen. Having shot a sleeping bird I smuggled it into camp in order not to offend the village folk. After having taken these precautions for about two months I learned that the people entertained no objection whatever to the birds being shot ! Peafowl are objects of veneration in all the Native States of Rajputana. These are strongholds of orthodox Hinduism. Nilgai, even, may not be shot, because the Pundits, not being zoological experts, labour under the delusion that these ungainly antelope are kine.

In some parts of India peafowl may be seen in a state of semi-domestication and are regularly fed by temple keepers. The drawback to the peacock as a domestic bird is that he renders the night hideous by his cries, which resemble those of an exceptionally lusty cat. Blanford, I notice, called them " sonorous." There is no accounting for taste. In my opinion, peafowl should be seen and not heard.

The peacock, like the ostrich, is almost omnivorous ; it feeds chiefly upon grain, buds, and shoots of grass, but it is not averse to insects, and will devour many of these, which are generally supposed to be inedible and so warningly coloured. Lizards and snakes complete a varied menu.

The peafowl is a bird of considerable interest to the zoologist, as it affords a striking example of sexual dimorphism. In plain English, the cock differs greatly from the hen in appearance. In some species, such as the myna, the crow, and the blue jay, the cock is indistinguishable from the hen. In others, as, for example, the sparrow, the sexes differ slightly in appearance. In others, again, the cock differs from the hen as the sun does from the moon. The peafowl is one of these.

Zoologists have for years been trying to find out why in some species the cock resembles the hen while in others it does not. Humiliating though it be, we must, if we are honest, admit that we are little, if any, nearer the explanation of the phenomenon than we were a couple of centuries ago. Darwin thought that the pretty plumage of the males was due to selection on the part of the females. He tried to prove that hens are able to pick and choose their mates, that they have a keen eye for beauty. Just as political economists of Ricardo's school teach us that every man marries the richest woman who will have him, so does Darwin ask us to believe that hens always mate with the best-looking of their suitors, that they quiz each with the eye of an art critic, and pronounce judgment somewhat as follows : " Number one is no class ; his train is too short. I would not be seen dead beside number two ; he looks as though he had issued from a fifth-rate dyer's shop. I'll take number three, he is the pick of the bunch." Darwin argued that the showy cocks are fully alive to their good looks, and know how to show them off to best advantage. There is much to be said in favour of his theory. A peacock, when he sees a hen that he admires, promptly turns his back upon her, erects his great train and his paltry little tail which is hidden away underneath. He then spreads out his feathers and suddenly faces the hen, flapping his wings, and causing every feather in his body to quiver with a curious noise, so that he appears to be seized with a shivering fit. The hen either affects not to notice him, or assumes an air of studied boredom. Unfortunately for Darwin's theory, peacocks sometimes show off in the absence of other living creatures. Moreover, a young cock with a train of which a magpie would be ashamed will strut about and show off with the greatest pride.

There are in the " Zoo " at Lahore a number of albino peacocks. These, although handsome birds, are not so beautiful as the coloured variety, being a uniform white ; nevertheless they are exceedingly popular with the hens, and experience no difficulty in cutting out all the coloured beaux. It is very naughty of the hens to prefer the albinos, for by so doing they deal a severe blow to the theory of sexual selection. Stolzman has quite another hypothesis to account for the superior beauty of the male. As any " suffragette " will tell you, the male is a more or less superfluous being ; the world would get along much better if he were less plentiful. Hence, in the interests of the race, it is necessary that the numbers of the pernicious creature should be strictly limited. Nature has, therefore, arrayed cock-birds in coats of many colours so that they shall be easily seen and devoured by beasts of prey ! Wallace, again, thinks that the comparatively sombre hues of the hen are due to her greater need of protection, as it is she who does all the incubating. An objection to this view is the well-known fact that many showy cocks sit on the eggs turn-about with the dull-coloured hens in open and exposed nests. Peafowl are polygamous. The breeding season begins in May and continues all through the hot weather. The typical nest is described as " a broad depression scratched by the hen, and lined with a few leaves and twigs or a little grass." It is usually made amongst thick grass or in dense bushes, but occasionally there is no attempt at concealment. Mr. A. Anderson states that peahens frequently lay at high elevations, that he has on several occasions taken their eggs from the roofs of huts of deserted villages on which rank vegetation grew to a height of two or three feet. My experience of captive birds bears out this. The peahens in the Lahore " Zoo " lay all their eggs on a broad shelf in their aviary, some fifteen feet above the level of the ground. Seven or eight eggs of a dirty white hue are laid. These are, in the words of Hume, " delicious eating."

XXIX

THE INDIAN REDSTART

POETS, naturalists, essayists, and novelists have with one accord and from time immemorial extolled the English spring. In this particular instance their eulogies are justified, for spring in England is like a wayward maiden : when she does choose to be amiable, she is so amiable that her past perverseness is at once forgiven. But why do not Anglo-Indian writers sing to the glories of the Indian autumn ? Is it not worthy of all praise ? It is the season which corresponds most nearly to spring in England, and is as much longed for. Even as spring chases away the gloomy, cheerless English winter, so does autumn drive away the Indian hot weather, unpleasant everywhere, and terrible in the plains of the Punjab and the United Provinces. Those condemned to live in Portland Gaol probably suffer fewer physical discomforts than they who spend the summer in any part of the plains of Northern India. First, weeks of a furnace-like heat, when to breathe seems an effort ; then a long spell of close, steamy heat, so that the earth seems to have become a great washhouse. From this the Anglo-Indian emerges, limp, listless, and languid.

How great, then, is his joy when one day he notices a suspicion of coolness in the air. Day by day this coolness grows more appreciable, so that by late September or early October to take an early-morning stroll becomes a pleasure. Then the sky is bluer, the atmosphere is clearer, the foliage is greener than at any other time of the year. Then at eventide the village smoke hangs low, looking like a thin blue semi-transparent cloud resting lightly on the earth—a sure sign of the approaching cold weather. Then, too, the winter birds begin to appear.

Even as the cuckoo is welcomed in England as the harbinger of the sweet spring, so in Northern India is the redstart looked for as the herald of the glorious cold weather. Within a week of the first sight of that sprightly little bird will come the day when punkahs cease to be a necessity. Last year (1907) the hot weather lingered long, and the redstarts were late in coming. It was not until the 27th September that I observed one at Lahore.

Several species of redstart are found within Indian limits, but only one of them haunts the plains, and so thoroughly deserves the name of the Indian redstart (*Ruticilla rufiventris*). This species visits India in hundreds of thousands from September to April.

I have observed it in the city of Madras, but so far south as that it is not common, being a mere straggler to those parts. In the Punjab and the United Provinces, however, it is exceedingly numerous. Throughout the cold weather at least one pair take up their abode in every compound.

M

The Indian redstart is a sexually dimorphic species, that is to say the cock differs from the hen in appearance ; the former, moreover, is seasonally dimorphic. The feathers of his head, neck, breast, and back are black with grey fringes. In the autumn and early winter the grey edges completely obliterate the black parts, so that the bird looks grey. But during the winter the grey edges gradually become worn away, and the black portions then show, so that by the middle of the summer the cock redstart is a black bird. Thus he remains until transformed by the autumnal moult. His under parts are deep orange, and his lower back and all the tail feathers, except the middle pair, are brick-red. Now, when the tail is unexpanded the two middle caudal feathers are folded over the others, and hide them from view, and, as the lower back is covered by the wings, the red parts are not visible when the bird walks about looking for food ; but the moment it takes to its wings all the red feathers become displayed, so that the bird, as it flies away, looks as though its plumage were almost entirely red. Hence the name redstart—" start " being an old English word for tail. Another popular name for the bird is firetail.

Two species of redstart visit England, and these also are characterised by reddish tails. The hen Indian redstart is reddish brown where the cock is grey or black, and red where he is red. The gradual change in colour undergone by the cock redstart every year is instructive, because it seems to show that the bird is even now undergoing evolution. I think it likely that the feathers of the cock were at one time uniformly grey and that they are becoming a uniform black. The tendency seems to be for the grey margin to become narrower. It will probably eventually disappear. In some birds it is so narrow that much black shows even after the autumn moult ; in others the margin is so broad that it never disappears. What is causing this change in plumage ? It cannot be the need for protection. The incipient blackness is probably an indirect result of either natural or sexual selection. Thus birds with black bases to their feathers may be either more robust or have stronger sexual instincts than those which have scarcely any black. In the former case natural selection, and in the latter sexual selection, will tend to preserve those individuals which have the least grey in their feathers. This idea of the connection between colour and strength is not mere fancy. Cuckoo-coloured (barred-grey) birds are very common among ordinary fowls, but are, I believe, never seen among Indian gamecocks. Grey plumage seems to be inconsistent with fighting propensities. Black, on the other hand, seems to be a good fighting colour. Most black-plumaged birds, as, for example, the king crow, the various members of the crow tribe, and the coot, are exceedingly pugnacious.

Redstarts live largely on the ground, from which they pick their food. This appears to consist exclusively of tiny insects. They sometimes hawk their quarry on the wing. They are usually found near a hedge or thicket, into which they take refuge when disturbed. They show but little fear of man, and, consequently, frequent gardens. They occasionally perch on the housetop. Indeed, they are quite robin-like in their habits, and the species, thanks to its reddish abdomen, looks more like the familiar English robin than does the Indian robin.

The Indian redstart, like all its family, has a peculiar quivering motion of its tail, which is especially noticeable immediately after it has alighted on a perch ; hence its Hindustani name, *Thir-thira*, the trembler. Its Telugu name is said to be *Nuni-budi-gadu*—the oil-bottle bird—a name of which I am unable to offer any explanation. Eurasian boys call it the " devil bird," for reasons best known to themselves.

The redstart stays in India until May, when it goes into Tibet and Afghanistan to breed. A few individuals are said to spend the summer in India. There are in the British Museum specimens supposed to have been shot at Sambhar in July and Ahmednagar in June. I have never observed this bird in India between the end of May and the beginning of September, and am inclined to think that the above dates have been incorrectly recorded.

XXX

THE NIGHT HERON

SOME American millionaires are said to sleep for only three hours out of the twenty-four. I do not believe this; I regard the story as a fabrication of the halfpenny paper. But, even if it be true, the night heron (*Nycticorax griseus*) is able to eclipse the performance. That bird only sleeps when it has nothing better to do. It looks upon sleep as a luxury, not a necessity. As its name implies, it is a creature of the night; but it is equally a day bird. You will never catch it napping. Just before sunset, when the crows, wearied by the iniquities they have wrought during the day, are wending their way to the corvine dormitory, the night herons sally forth from the trees ("roosts" would be a misnomer for them) in which they have spent the day and betake themselves, in twos and threes, to the water's edge. As they fly they make the welkin ring with their cries of *waak, waak,* or *quaal, quaal*—sounds which may be likened to the quacking of a distressed duck. Having arrived at their feeding-ground, they separate and proceed to catch fish and frogs in the manner of the orthodox heron. After an all-night sitting, or rather standing, in shallow water, they return to their day quarters, where they are popularly supposed to sleep. They may possibly spend the day in slumber when they have neither nests to build nor young to feed. I am not in a position to deny this, never having visited a heronry on such an occasion. I speak, however, as one having authority when I say that all through the nesting season the night heron works harder during the hours of daylight than the British workman does. At the present time (April) thirty or forty night herons are engaged in nesting operations in the tall trees that grow on the islands in the ornamental pond that graces the Lahore Zoological Gardens, and as I visit those gardens almost daily I have had some opportunity of observing the behaviour of our night bird during the daytime. I may here say that night herons seem very partial to Zoological Gardens, inasmuch as they also resort to the Calcutta " Zoo " for nesting purposes. This is, of course, as it should be. Every well-conditioned bird should bring up its family in a " zoo " by preference. Had birds the sense to understand this, many of them would be spared the miseries of captivity.

Before discoursing upon its nesting habits it is fitting that I should try to describe the night heron, so that the bird may be recognised when next seen. I presume that everyone knows what a heron looks like, but possibly there exist persons who would be at a loss to say wherein it differs from a stork or a crane. It may be readily distinguished from the latter by its well-developed hind toe. Storks and herons are

perching waders, while cranes do not trust themselves
to trees because they cannot perch, having no hind
toe to grasp with. The heron's bill is flatter and more
dagger-shaped than that of a stork. Moreover the
former possesses, inside the middle claw, a little comb,
which the stork lacks. The heron flies with neck drawn
in, head pressed against the back, and beak pointing
forwards. It never sails in the air, but progresses, like
the flying-fox, with a steady, continuous flapping
motion. So much for herons in general. To those who
would learn more of these and other long-legged fowls
I commend Mr. Frank Finn's excellent little book,
entitled *How to Know the Indian Waders*.

The night heron is considerably smaller than the
common heron—the heron we see in England, and
larger than the Indian paddy bird—the ubiquitous
fowl that looks brown when it is standing and white
when it is flying. The head and back of the night heron
are black, the remainder of the upper plumage is grey,
the lower parts are white. There are two or three long,
white, narrow feathers, which grow from the back of
the head and hang down like a pigtail. The eye is
rich ruby-red. Young night herons are brown with
yellowish spots, and the eye is deep yellow.

Any resident of Madras may see this species if he
repair to the Redhills Tank. One of the islands in that
tank supports a considerable population of night herons
and little cormorants. The former nest in the trees
on the island in July. The place is well worth visiting
then. As the boat carrying a human being approaches
the islet, all the night herons fly away without a sound.

They love their young, but not so much as they love themselves, so they leave their offspring at the first approach of the human visitor and remain away until he turns his back on their nesting-ground. A night heron never allows his valour to get the better of his discretion. The nest is a platform of twigs placed anywhere in a tree. Four pale greenish-blue eggs are laid. A heronry is a filthy place. The possessors are, like our Indian brethren, utterly regardless of the principles of sanitation. The whole island will be found white with the droppings of the birds, and the unsavoury smell that emanates therefrom would do credit to a village inhabited by *chamars*. Although they are evil-tempered, cantankerous creatures, night herons always nest in company. It is no uncommon thing to find half a dozen nests in the same tree, so that the sitting birds are able to compare notes while engaged in the duties of incubation. Both the parent birds take part in nest construction, and, as they work by day, it is quite easy to watch the process. They wrench small branches from trees, and, as they have only the beak with which to grasp these, they find twig-gathering hard work. When a twig has been secured it is dropped on to the particular part of the tree in which the bird has thought fit to build. Forty or fifty twigs dropped haphazard in a heap constitute the nest. The birds make a great noise while engaged in building. Quarrels are of frequent occurrence. It sometimes happens that two birds want the same twig; this invariably gives rise to noisy altercation. The crows too are provocative of much bad language on the part of the night herons. Whenever any of the crows of the neighbourhood has nothing else to do, he says to a kindred spirit : " Come, let us worry the night herons." Whereupon the pair—*Arcades ambo*—go and pretend to show the herons how to build a nest.

When, my friends, you consider the untidiness and filthiness of the heron's nest, you will be able to appreciate to the full the audacity of the latest falsehood circulated by the plume trade—to wit, the egret plucks out its nuptial plumes, which constitute the " osprey " of commerce, and weaves these into the nest to make it more cosy ; and, after the young ones are fledged, some honest fellow visits the nest and disentangles the plumes therefrom !

A baby heron is a disgustingly ugly creature. It is a living caricature. Patches of long hair-like feathers are studded, apparently haphazard, over its otherwise naked body and give it an indescribably grotesque appearance. It looks like a bird in its dotage. If you lift a young heron out of the nest you will probably find that his " corporation " is distended to bursting-point, and, if you do not handle him carefully, a half-digested frog will, as likely as not, drop out of him !

The farther north one goes the earlier in the year does the night heron breed. In Kashmir the nesting season is in full swing in March. In the Punjab April and May are the nesting months ; in Madras the birds do not begin to build until July ; and I have seen eggs at the end of August. It is my belief that the night heron is a migratory bird. During the winter months not a single specimen of that species is to be seen in or about Lahore, but they migrate there regularly every April. They disappear again to I know not where when they have reared up their young.

XXXI

THE CEMENT OF BIRD MASONS

BIRDS may be divided into two classes—those which build nests and those which do not. To the latter belong the parasitic starlings and cuckoos, which drop their eggs in the nests of other birds ; those, such as plovers, which lay their eggs on the bare ground ; and those which deposit them in holes, in the earth, in trees, in banks, or in buildings, as, for example, the Indian roller or blue jay (*Coracias indica*).

Intermediate between the birds that build nests and those which do not—for there are no sudden transitions, no sharply defined lines of division in nature—are those birds which merely furnish, more or less cosily, the ready-made holes in which they deposit their eggs. The common myna (*Acridotheres tristis*) affords a familiar instance of this class of birds. Some of the nest-builders are really excavators ; they dig out their nests in a tree or bank. The woodpeckers and the bee-eaters are examples of these. The rest of the nest-builders actually construct their nurseries. These buildings are of various degrees of complexity. Crows, doves, birds of prey, herons, and a few other families are content with a mere platform of sticks and twigs, which rests in the fork of a tree, or on a ledge or other suitable surface. The birds which build primitive nests of this description are not put to the trouble of seeking or manufacturing any cohesive materials. It is only when the nest takes some definite shape and form that means have to be found of binding together the materials of which it is composed, and of attaching the whole to that which supports the nest. In such cases the component materials are either woven or cemented together. It is among the woven nests that we find the highest examples of avian architecture. The homes of the weaver-bird (*Ploceus baya*) and of the Indian wren-warbler (*Prinia inornata*) are constructed with a skill that defies competition. But it is not with these wonderful nests that we are concerned to-day. It must suffice to say that woven nests have to be supported ; they cannot float in the air. There are various methods of supporting them. The nest may be firmly wedged into a forked branch. It may be bound to its supports, as in the case of the nest of the king crow (*Dicrurus ater*). The supporting branches may be worked into its structure, as is done by *Prinia inornata*. The nest may hang, as does that of *Ploceus baya*. It may be cemented to its support, as in the case of the nests of the various swifts ; or it may rest on supporting fibres which are slung on to a forked branch, just as a prawn net is slung on to its frame. The golden oriole (*Oriolus kundoo*) resorts to this ingenious device.

Coming now to those nurseries in which the building

materials are cemented together, we must first consider the nests of the swallows and swifts. These birds secrete a very sticky saliva, which quickly hardens when it is exposed to the air. This constitutes an excellent cement. Watch a swift working at its nest under the eaves of a house. It flies to it with a feather or piece of straw carried far back in the angle of its mouth, hangs itself by means of its four forwardly directed toes on to the half-completed nest, which is stuck on to the wall of the house, and, having carefully placed the feather or straw in the required position, holds it there until the sticky saliva it has poured over it has had time to harden and thus firmly glue the added piece of material to the nest. The bits of straw, feathers, etc., may be said to constitute the bricks, and the saliva the cement of the swift's nest. Some swifts build their nests exclusively of their saliva. These constitute the " edible birds' nests " of commerce, and may be likened to houses built entirely of cement. The martin (*Chelidon urbica*), the common swallow (*Hirundo rustica*), and the wire-tailed swallow (*H. smithii*) construct their nests of clay and saliva. They repair to some puddle and there gather moist clay, which they stick on to some building, so as to form a projecting saucer-shaped shelf. In this the eggs are laid. But nature has not vouchsafed sticky saliva to all birds, so that many of them have to find their cement just as they have to seek out the other building materials they use.

The chestnut-bellied nuthatch (*Sitta castaneiventris*), which nestles in holes in trees, fills up all but a small part of the entrance with mud " consolidated with some viscid seed of a parasitical plant."

The hornbills close up the greater part of the orifice of the hole in which they nest with their droppings mixed with a little earth.

Hume informs us the rufous-fronted wren-warbler (*Franklinia buchanani*) utilises a fungus as its cement. " In all the nests that I have seen," he writes, " the egg-cavity has been lined with something very soft. In many of the nests the lining is composed of soft, felt-like pieces of some dull salmon-coloured fungus, with which the whole interior is closely plastered."

The cement which is most commonly used is cobweb. I do not think that I am exaggerating when I say that cobweb enters extensively into the structure of the nests of more than one hundred species of Indian birds. What birds would do without our friend the spider I cannot imagine.

The nest of some birds is literally a house of cobwebs. The beautiful white-browed fan-tail fly-catcher (*Rhipidura albifrontata*) is a case in point. Its nursery is so thickly plastered with cobweb as to sometimes look quite white. It is a tiny cup that rests on a branch of a bush or small tree, and is composed of fine twigs and roots, which are cemented to the supporting branch and to one another by cobweb. This the bird takes from the webs of those trap-door spiders which weave large nets on the ground.

Utterly regardless of the feelings of the possessor of the web, the fly-catcher takes beakful after beakful of it, and smears it over the part of the branch on which the nest will rest. It then sticks to this some dried grass stems or other fine material, next adds more cobweb, and continues in this manner until the neat little cup-shaped nest is completed. This, as I have already said, is thickly coated exteriorly with cobweb to give it additional strength.

The sunbirds or honeysuckers make nearly as extensive use of cobweb in nest construction as do the fan-tailed fly-catchers. Loten's honeysucker (*Arachnechthra lotenia*) seeks until it finds a large spider's web stretched horizontally across some bush; it then proceeds to build its nursery in the middle of this. As the material is added the nest grows heavier, and thus depresses the middle of the web until it at last assumes the shape of a V, in the angle of which the mango-shaped nest is situated. The nursery is thus suspended from the bush by the four corners of the cobweb.

A spider's web looks such a flimsy affair that it does not seem possible that it could support a nest peopled by a number of birds. Sometimes the nest derives additional support by being attached to other branches. Moreover, a tiny creature such as a sunbird is almost as light as the proverbial feather. Then cobweb is exceedingly elastic, and, considering its attenuity, is able to support a surprising amount of weight. It occasionally happens that the common garden spider (*Epeira diadema*) is not able to find a *point d'appui* to which it can attach the lower part of its web; it then utilises a stone (which may be as much as a quarter-inch in each dimension) as a plummet to make the nest taut. This comparatively heavy stone hangs by a single thread.

I have sometimes amused myself by testing the strength of a strand of cobweb stretched across a path, by hanging bits of match or other light material on it. In one experiment a gossamer thread, thirty feet in length, stretched across a road, bore the weight of five blades of grass which were hung upon it. The sixth blade proved to be the last straw that broke the camel's back.

The strength of cobweb is proved by the fact that many of the birds that build hanging nests use it as cement to attach them to the supports from which they are suspended. The Indian white-eye (*Zosterops palbebrosa*) fixes its tiny oriole-like nest to the supporting branches, not by fibres, but by cobweb. In the same way the yellow-eyed babbler (*Pyctorhis sinensis*), whose nest is shaped like an inverted cone, attaches this by cobweb to the stems of the crop in which it is situated.

The common honeysucker (*Arachnechthra asiatica*), whose nest looks like a tangle of dried twigs and other rubbish, uses much cobweb in the construction thereof. The little nursery is suspended by means of cobweb from some projecting branch of a bush, and the various materials which compose it are stuck together with spider's web ; but in this case some sticky resinous substance is usually used in addition to the cobweb.

The tailor-bird (*Orthotomous sutorius*) always uses cobweb to draw together the edges of the leaf or leaves that compose its nest. Having made a series of punctures along the edges of the leaf to be utilised, it procures some cobweb, and, having attached it to one edge of the leaf, carries the strand across to the other edge and, before attaching it to this, pulls it so tightly as to draw the two edges together. When the nest has taken its final shape the bird strengthens the first attenuated strands of cobweb by adding more cobweb or some threads of cotton.

Many birds which weave their nests plaster the exterior more or less thickly with cobweb so as to add strength to the structure.

It would be wearisome to detail all the kinds of nest into the composition of which cobweb enters. Sufficient has been said to show that this very useful substance is the favourite cement of bird masons.

N

XXXII

INDIAN FLY-CATCHERS

THERE exist in the Indian Empire no fewer than fifty-one species of fly-catcher. This fact speaks volumes for the wealth of both the bird and the insect population of India. Fly-catchers are little birds that feed exclusively on insects, which they secure on the wing. Their habit is to take up a strategic position on some perch, usually the bare branch of a tree, whence they make sallies into the air after their quarry. Having secured the object of their sortie—and this they never fail to do—they return to their perch. A fly-catcher will sometimes make over a hundred of these little flights in the course of an hour ; the appetite of an insectivorous bird appears to be insatiable. All fly-catchers obtain their food in this manner, but all birds which behave thus are not members of the fly-catcher family. As fly-catchers are characterised by rather weak legs, and, in consequence, do not often descend to the ground, they are of necessity confined to parts of the country well supplied with trees. Thus it comes to pass that the great majority of fly-catchers are found only in well-wooded hill tracts. Four or five species, however, occur commonly in the plains. With two of these—the glorious paradise fly - catchers (*Terpsiphone*) and the very elegant fan-tail fly-catchers (*Rhipidura*)—I have dealt in my former books. I therefore propose to confine myself to some of the many other species. Of these last, the brown fly - catcher (*Alseonax latirostris*) is the one most frequently met with in the plains. This is the most inornate of all the fly-catchers. As its name implies, brown is its prevailing hue. Its lower parts are, indeed, whitish, and there is an inconspicuous ring of white feathers round the eye, but everything else about it is earthy brown. It is the kind of bird the casual observer is likely to pass over, or, if he does happen to observe it, he probably sets it down as one of the scores of warblers that visit India in the cold weather. It is only when the bird makes a sudden dash into the air after an insect that one realises that it is a fly-catcher. The brown fly-catcher is an Ishmaelite. It seems never to remain for long in one place, and, although it may be seen at all times of the year, its nest does not appear ever to have been found in this country.

A more ornamental fly-catcher which occasionally visits the plains is the grey-headed fly-catcher (*Culicapa ceylonensis*). In this species the head, neck, and breast are ash-coloured, the wings and tail are dark brown, the back greenish yellow, and the lower parts dull yellow. This fly-catcher is common both in the Nilgiris and the Himalayas. It has the usual habits of the family. Like the majority of them it is no songster, although it frequently emits a cheeping note. Its nest is a very beautiful structure, a ball of moss which is attached to a moss-covered tree or rock, more often than not near a mountain stream.

Fly-catchers usually nidificate in the neighbourhood of water, because that element favours the existence of their insect food.

Siphia parva—the European red-breasted fly-catcher—is a species which visits the plains of India in the cold weather, but not many individuals penetrate so far south as Madras. This bird is easily recognised, since the cock bears a strong likeness to the familiar English robin red-breast. I may here mention that an allied species—the Indian red-breasted fly-catcher, *S. hyperythra*—summers in Kashmir and winters in Ceylon, but, curiously enough, it has not been recorded from the plains of India. It would thus seem to fly from Kashmir to Ceylon in a single night. Even so, it would be very extraordinary if an occasional individual did not fail to perform the whole journey in so short a space of time ; therefore, this species should be watched for in South India in spring and autumn. It is easily distinguished from allied species by a black band which surrounds the red breast and abdomen.

As it is impossible to detail in one brief essay all the species of fly-catcher found in the Indian hills, I propose merely to mention those that are most common in the Nilgiris and the Himalayas, and then to make a few observations on fly-catchers in general. In addition to the fan-tail, the grey-headed and the brown fly-catchers,

the following species are abundant in the Nilgiris :
Tickell's blue fly-catcher (*Cyornis tickelli*), the Nilgiri
blue fly-catcher (*Stoparola albicaudata*), and the black
and orange fly-catcher (*Ochromela nigriruja*). In the
Himalayas, the paradise fly-catcher is common in
summer at lower altitudes. Above 6000 feet elevation
the following are the species most commonly seen :
the grey-headed fly-catcher, the white-browed blue
fly-catcher (*Cyornis superciliaris*), and the beautiful
verditer fly-catcher (*Stoparola melanops*), which is no
mean songster.

Fly-catchers form a most interesting group of
birds. It is, I maintain, quite impossible for any
man possessed of a logical mind to contemplate
this family without discovering that the theory
of natural selection is utterly inadequate to account
for the variety of animal life that exists upon the
earth. The habits of practically all the fly-catchers
are identical. They all dwell in an arboreal habitat ;
nevertheless, the various species display great dis-
similarity in outward appearance. Some species are
brightly plumaged, others are as dully clad as a bird
can possibly be. Some have crests and long tails, others
lack these ornaments. The adult cock paradise fly-
catcher, with his long, white, satin-like tail feathers,
is the most striking of birds, while the brown fly-
catcher is less conspicuously attired than a hen sparrow.
This is not the only difficulty presented to the theory
of natural selection by fly-catchers. In some species,
as, for example, the paradise fly-catcher, the sexes are
altogether dissimilar in appearance, while in others the

most practised eye cannot distinguish between the
cock and the hen. Nor does there appear to be any connection between nesting habits and the presence or
absence of sexual dimorphism. The fan-tail fly-
catchers, in which the sexes are alike, and the paradise
fly-catchers, in which they differ widely, both build
little cup-shaped nests in the lower branches of trees,
and in both the cock shares with the hen the duty of
incubation. Again, the verditer and the white-browed
blue fly-catchers build their nests in holes in trees ;
yet in the former both sexes are blue, while in the
latter the cock only is blue.

Further, in the fly-catchers we see every gradation
of sexual dimorphism, from a difference so slight
as to be perceptible only when the sexes are seen side
by side, to a difference so great as to make it difficult
to believe that the sexes belong to one and the same
species. It must, therefore, be obvious to any sane
person that neither natural nor sexual selection can be
directly responsible for the colouration of many species
of fly-catcher.

Another interesting characteristic of the fly-catchers
is the total absence of green in the plumage of any
of them. They are birds of a variety of colours ;
they display many shades of blue, yellow, orange, red,
grey, and brown, also black and white ; but not one
carries any green feathers. Yet they are essentially
arboreal birds, so that green would be a very useful
colour to them from the point of view of protection from
enemies. From the fact, then, that none of the fly-
catchers are green, we seem to be compelled to infer

that there is something in their constitution that prevents green variations appearing in their plumage.

In conclusion, note must be made of the fact that fly-catchers, although they subsist almost entirely upon insect diet, appear but rarely to devour butterflies. I have watched fly-catchers closely for several years, and have on two occasions only seen them chase butterflies or moths. Five years ago in Madras I observed a paradise fly-catcher chasing a small butterfly, and recently, in the Himalayas, I saw a grey-headed fly-catcher drop down from a tree and seize a moth that was resting in the gutter. The reason why fly-catchers do not often attack butterflies is obvious; these insects offer very little meat and a great deal of indigestible wing surface. Nevertheless, the theory of protective mimicry is almost exclusively illustrated by examples taken from butterflies. In theory, these creatures are so relentlessly persecuted by insectivorous birds that in order to escape their foes many edible butterflies mimic the appearance of unpalatable species. Unfortunately for theory, few creatures in practice seem to attack butterflies when on the wing, which is just the time when the " mimicry " is most obvious.

The elegant little fly-catchers, then, are birds which mock Darwin, laugh at Wallace, and make merry at the expense of Muller and Bates !

XXXIII

INSECT HUNTERS

FLY-CATCHERS, although they subsist almost entirely on insects, are by no means the only insectivorous creatures in existence. They merely form a considerable branch of the Noble Society of Insect Hunters.

If there exist any philosophers in the insect world they must find the uncertainty of life a fitting theme on which to lavish their philosophical rhetoric. Consider for a moment the precariousness of the life of an insect! There exist in India probably over three hundred species of birds which live almost exclusively upon an insect diet. Think of the mortality among insects caused by these birds alone, by the mynas, the swifts, the bee-eaters, the king crows, *et hoc genus omne*. Then there are insectivorous mammals, to say nothing of man who yearly destroys millions of injurious and parasitic hexapods. Fish too are very partial to insects, while for spiders, frogs, and lizards, life without insects would be impossible. Nor do the troubles of insects end here, they are preyed upon by their own kind, and, strange phenomenon, some plants entrap and destroy them. But we Anglo-Indians cannot afford to sympathise with the insects. In spite of the high mortality of the hexapod tribes, they flourish like the green bay-tree. So prolific are they that, notwithstanding the fact that millions are daily destroyed by their foes, the life of human beings in India becomes a burden on account of the creeping things. In the monsoon the insects tax man almost to the limits of his endurance—they teaze, bite, and worry his person, they destroy his worldly goods, and, not content with this, find their way into his food and drink. For this reason I feel very kindly disposed to the frogs, the lizards, and the fly-catching birds.

It is worth coming to India if only to see a frog or toad at work. Go at sunset, during a break in the rains, on to the *chabutra*, and place a lamp near you. Thousands of insects of all shapes and sizes are attracted by the light. In their wake come the toads. A toad always looks *blasé*. His stupid appearance and sluggish movements give him this air. Watch him as he hops into the zone of light. He advances to within an inch of a resting insect, and, before you can say " Jack Robinson," the creature has flown into his mouth ! The toad takes another hop, and a second insect follows the example of the first ; then another and another ! Have the insects all suddenly gone mad ? Are they bewitched, mesmerised by the ugly face of the toad ? Nothing of the kind. The insects have not jumped into the amphibian's mouth at all. The toad has a long tongue attached at the front end to its mouth. This tongue is covered with sticky saliva and is capable of being protruded and retracted with lightning rapidity. In other words, the toad's tongue is just a fly-paper, capable of the most perfect manipulation. The unsuspecting insect is resting, and hears not the silent approach of its enemy. Suddenly it is caught up by a great sticky tentacle, then comes oblivion. The toad's tongue has shot forth and back again so quickly as to be imperceptible to the human eye.

The lizard obtains its food in a similar way. It enters the bungalow and lies up during the day behind a picture. As soon as the lamps are lighted it comes forth as hungry as the proverbial hunter. In a single night it devours hundreds of insects. I have watched a lizard feeding in this way until he had consumed so many insects that he could scarcely move : and doubtless he would have continued his gluttonous meal but for the fact that he had become as slow as Mark Twain's jumping frog after it had partaken copiously of shot ! The lizard cannot shoot out his tongue to the extent that the frog can, so he has to make a dash at each insect before swallowing, and, to his credit, it must be said that he rarely lets a victim escape him unless, of course, he has over-eaten himself.

Although I am very fond of the nimble little gecko, I must admit that he is an out-and-out glutton. Sometimes his gluttony leads him to try to capture quarry beyond his capacity. Let me relate an amusing little incident that I recently witnessed. The scene was my dressing-table, and the time 9 p.m. in the month of August ; the day I forget. It matters not. A large stag-beetle was crawling laboriously across the dressing-table. Upon this table was an ordinary looking-glass, under which a lizard had taken up his habitation. From his point of view the position was a good one, for the lamp overhead attracted to the table a number of insects which the lizard could watch from under the base of the glass.

The lizard caught sight of the beetle and began to stalk it. Surely, I thought, the lizard will not try to devour that beetle, which is nearly half as big as himself ; but, as he emerged from under the glass, I saw that he meant business. Slowly but surely he gained upon the slow-moving beetle. Having arrived close up behind it, he shot forth his sticky tongue. The next moment the beetle found itself lying on a spot eight inches from where it had a second before stood, and the lizard was trembling in his lair. The reptile had apparently expected to find the beetle as soft and luscious as a strawberry, so the instant his tongue felt the hard, chitinous integument of the beetle he drew that organ back pretty smartly. But his tongue was so sticky that the beetle stuck to it for a moment, and so was thrown backwards over the reptile's head. The lizard was startled at what had happened, so instinctively took cover. The insect too was scared nearly out of its wits, and did what most frightened insects do, that is to say, retracted its legs and remained perfectly motionless. When, however, several minutes passed and nothing happened, the beetle grew bold, and putting forth its legs, began again to crawl on its way. Directly it moved the lizard put himself on the *qui vive*, and even went so far as again to follow it, but, profiting by his recent experience, did not attempt a second time to swallow it. Thus the beetle passed off the stage.

Seeing that this particular lizard was not over sharp, I determined to play a little practical joke upon it. Taking a piece of black worsted, I rolled it up into a ball about the size of a fine, strapping blue-bottle fly, and, having attached a piece of cotton to it, I dangled this bait before the lizard. I succeeded in " drawing " him. He was on it before I could say " knife."

In less than a second the worsted was in his mouth, but he dropped it like a hot potato, and then sulked under the looking-glass, apparently greatly annoyed at having been made a fool of twice in succession. The next day I chanced to come upon a toad, busy catching insects. Wondering whether he would be deceived, I threw on to the grass near him the end of a lighted cigarette which I had been smoking. He at once caught sight of it, and sat there looking at it intently for some seconds, and I began to think he would not fall into the trap, but the temptation was too strong, for he shot forth his tongue to seize it. He discovered that the " tongue is an unruly member " as he re-tracted the smarting organ.

It is therefore clear that some insect-hunters are ever ready to try experiments as regards food.

Fish too, when really hungry, do not appear to exercise much discrimination as to the nature of the " fly " they will take.

The swarming of the " white ants " is a red-letter day

for the insect-eating animals, an annual harvest in which they revel. The mynas and the crows do not disdain to partake of this copious meal supplied by nature.

The latter are omnivorous birds ; all is grist which comes to their mill—carrion, fruit, locusts, termites, fish, grain, and the crumbs which fall from man's table.

The mynas too eat a variety of food, but they are first and foremost insectivorous birds. They are never so happy as when chasing grasshoppers on the grass. By preference they accompany cattle, strutting along beside these and catching in their beaks the insects as these latter jump into the air, frightened by the approach of the great quadruped.

The beautiful white cattle egrets (*Bubulcus coromandus*) in a similar way make buffaloes and kine act as their beaters.

The familiar king crow (*Dicrurus ater*) adopts two methods of insect-catching. The one he favours most is that of the fly-catcher. Sometimes, however, he attaches himself to a flock of mynas. In such cases he flies to the van of the flock and squats on the ground, regardless of the fact that by so doing his beautiful forked tail gets dusty. As the mynas approach, snatching up grasshoppers, they put up a number of flying insects, and these the king crow secures on the wing. As soon as the last of the mynas has passed by the king crow again flies to the van and repeats the performance.

In India almost every company of mynas has its

attendant king crow. Usually the two species are on good terms, but sometimes the king crow gets " above himself," and then there is trouble. The other day I saw a bank myna (*Acridotheres fuscus*) hop on to a king crow's back and administer unto him chastisement in the shape of a couple of vigorous pecks on the back of the head. On being released the king crow did not attempt to retaliate, but flew meekly away.

Among the *élite* of the insect-hunters we must number the swifts. Strange birds are these. Not once in their lives do they set foot upon the ground. For hours at a time they pursue their speedy course through the thin air, snatching up, as they move at full speed, minute insects.

But even their powerful pinions cannot vibrate for ever, so at intervals they betake themselves to the verandah of some bungalow, and there hang on to the wall close under the roof. Their claws are simply hooks, and this is their rest—clinging to a smooth horizontal wall !

So long is the list of insect-hunters, and so varied are their methods, that I am unable to so much as mention many of them. I must content myself, in conclusion, with noticing the tits, cuckoo-shrikes, minivets, and white-eyes, which flit from leaf to leaf, picking up tiny insects ; babblers and laughing thrushes, which spend the day rummaging among fallen leaves for insects ; nuthatches and tree-creepers, which run up and down tree-trunks on the hunt for insects ; and woodpeckers, which seize, by means of their sticky tongue, the insects they have, by a series of vigorous taps, frightened from their hiding-places in the bark.

Consider these, and you cannot but be impressed with the trials and troubles of an insect's life!

XXXIV

THE ROSY STARLING

EVERY Anglo-Indian is acquainted with the rose-coloured starling (*Pastor roseus*), although some may not know what to call it. Nevertheless, it is a bird of many aliases; to wit, the rosy pastor, the *tillyer*, the *cholum* bird, the *jowaree* bird, the mulberry bird, the locust-eater, the *golabi maina*. The head, neck, breast, wings, and tail are glossy black, while the remainder of the plumage is a pale salmon or faint rose-colour. The older the bird the more rosy it becomes, but the great majority are pale salmon, rather than pink.

Rose-coloured starlings are sociable birds. They go about in large companies, which sometimes number several thousand individuals. They are cold-weather visitors to India, spreading themselves all over the peninsula, being most abundant in the Deccan. In the north straggling flocks occur throughout the winter, but it is in April that they are seen in their thousands, preparatory to leaving the country for breeding purposes. These great gatherings tarry for a short time in Northern India while the mulberries and various grain crops are ripening. They seem to subsist chiefly upon these, whence some of their popular names, and the malice which the farmer bears them. They are undoubtedly a very great scourge to the latter, but they are not an unmixed pest, for they are said to devour locusts with avidity when the opportunity presents itself. Now, the slaying of a locust is a work of merit which ought to neutralise a multitude of sins.

The rosy starlings which occur in India are said to nest in Asia Minor. This may be so, but I am inclined to think that there must be some breeding-grounds nearer at hand, for these birds have been observed in India as late as July, and they are back with us again in September. To travel to Asia Minor, construct nests, lay eggs, hatch these out, rear up the young, and return to India with them, all within the space of two months, is an almost impossible feat. It is, of course, probable that the birds which remain in India so late as July do not return as early as September.

The large flocks of rosy starlings are quite a feature of spring in Northern India. On the principle that many hands make light work, a company of these birds experiences no difficulty in speedily thinning a crop of ripening corn. The starlings feed chiefly in the morning and before sunset. During the heat of the day they usually take a long rest, a habit for which the crop-watchers ought to be very thankful. When not feeding, rosy starlings usually congregate in hundreds in lofty trees which are almost bare of foliage. They then look like dried leaves. I have spoken of this as a rest, which is not strictly accurate.

o

They certainly do not feed, but they constantly flit
about from branch to branch, and do a great deal of
feather preening, and, during the whole day, they
give forth a joyful noise. Their note is a sibilant twitter
which is not very loud ; indeed, considering the efforts
put into it, there is remarkably little result, but the
notes are so persistent, and so many birds talk at once,
that they can be heard from afar. The song of the
rosy starling is not musical, not more so than the
" chitter, chitter " of a flock of sparrows at bed-time,
yet it is not displeasing to the ear. There is an exuber-
ance in it which is most attractive. It cannot be
conversational, for all the birds talk at once, and their
notes lack expression and variety. Their clamour is
not unlike the singing of the kettle as it stands on the
hob ; in each case the sound is caused by the letting
off of superfluous energy. Starlings literally bubble
over with animal spirits. There can be no question
as to their enjoyment of life.

Rosy starlings are the favourite game birds of the
natives of Northern India, for they are very good to eat
and easy to shoot. When a thousand of them are
perched in a bare tree, a shot fired into " the brown "
usually secures a number of victims. It is, therefore,
not difficult to obtain a big bag. Needless to say, the
natives shoot these birds sitting. The way in which
Europeans persist in firing only at flying objects is
utterly incomprehensible to the average Indian ; he
regards it as part of the magnificent madness which is
the mark of every sahib. I once asked a native *Shikari*
if he had ever fired at a flying bird. He was a gruff

old man, and not afraid to express his feelings. He looked me up and down with eyes filled with withering contempt, and said " What do you take me for ? Am I a sahib, that I should waste powder and shot on flying things ? I never fire unless I think that by so doing I am likely to bring down at least six birds."

It is impossible to watch a flock of *jowaree* birds without being struck by what I may perhaps term their corporate action, the manner in which they act in unison, as though they were well-drilled soldiers obeying the commands of their officer. This phenomenon is observable in most species of sociable birds, but, so far as I am aware, no ornithologist, save Mr. Edmund Selous, has paid much attention to the matter, or attempted to explain it. To illustrate. A flock of rosy starlings will be sitting motionless in a tree giving vent to their twittering notes, when suddenly, without any apparent cause, the whole flock will take to its wings simultaneously, as if actuated by one motive, nay, as if it were one composite individual. Again, a flock will be moving along at great speed, when suddenly the whole company will make a half-turn, and continue the flight in another direction. Yet again, a number of rosy starlings will be speeding through the air when six or seven of them, suddenly and simultaneously, change the direction of their flight, and thus form, as it were, a cross current. How are we to explain these simultaneous changes of purpose ? It is not, at any rate, not always, a case of " follow my leader," for frequently no one individual moves before the others. In some cases at least the change in purpose is not due to any command, no sound being uttered previous to one of these sudden impulsive acts. Mr. Selous seeks to explain the phenomenon by assuming that " birds, when gathered together in large numbers, act, not individually, but collectively, or rather, that they do both one and the other." According to him, the simultaneous acts in these cases are the result of thought-transference—a thought-wave passes through the whole flock.

Some may be inclined to scoff at this theory, but such will, I think, find it difficult to put forward any other explanation of the difficulty. As Mr. Selous points out, it seems " a little curious that language of a more perfect kind than animals use has been so late in developing itself, but animals would feel less the want of a language if thought-transference existed amongst them to any appreciable extent." Whether Mr. Selous has hit upon the correct explanation I hesitate to say. There is, however, no denying the fact that flocks of birds frequently act with what he calls " multitudinous oneness."

XXXV

THE PIED STARLING

WRITING of pied starlings (*Sturnopastor contra*) Colonel Cunningham thus delivers himself : " They are not nearly such attractive birds as the common mynas, for their colouring is coarsely laid on in a way that recalls that of certain of the ornithological inmates of a Noah's Ark ; their heads have a debased look, and they have neither the pleasant notes nor the alluringly familiar ways of their relatives." The above statement is, in my opinion, nothing short of libel. There are few living things more charming than pied mynas. These birds are clothed in black and white. Now a black and white garment usually looks well whether worn by a human being or an animal. In the case of the pied myna, or *ablak* as the Indians call it, the black and white are tastefully arranged. The head, neck, upper breast, back, and tail are glossy black, save for a large white patch on the cheek, which extends as a narrow line to the nape, a white oblique wing bar, and a white rump. The lower parts are greyish white. The bill is yellow, of deeper hue at the base than at the tip. I fail

to see in what way the head of the pied starling
has a debased look ; it is typical of its family. The
bill, however, is a trifle longer and more slender than
that of the common myna. The statement that pied
mynas have not the pleasant notes of the common
species is the most astounding of a series of astounding
assertions ; as well might a musician complain that the
cathedral organ lacks the fine tones of the street hurdy-
gurdy ! I like the cheerful " kok, kok, kok, kekky,
kekky " of the common myna. I also enjoy listening
to the harsh cries with which he greets a foe. India
would be a duller country than it is without these
familiar sounds, but I maintain that his most ardent
admirer can scarcely believe the common myna to be
a fine songster. The notes of the pied starling, on the
other hand, although essentially myna-like, are really
musical. Its lay is that of *Acridotheres tristis*, purified
of all the harshness, with an added touch of melody.
Jerdon, I am glad to notice, speaks of its pleasant
song, and Finn, who knows the bird well, writes in one
place of its beautiful note, and in another says : " It
does not indulge in any set song apparently, but its
voice is very sweet and flute-like, and it appears not
to have any unpleasant notes whatever—a remarkable
peculiarity in any bird, and especially in one of this
family." In Northern India the cheerful melody of
the pied starlings is one of the most pleasing adjuncts
of the countryside.

So jovial a bird is *Sturnopastor contra* that it is a
great pity that his range is comparatively restricted.
He would be a great acquisition to Madras and Bom-

bay. Unfortunately, the species is not found in South India, and is almost unknown in the Punjab. Agra is the most westerly place in which I have seen pied mynas. In Burma the species is replaced by an allied form, *S. superciliaris*, readily distinguished by the possession of a white eyebrow. By the way, I should be very glad if our Wallaceian friends would tell us why it is necessary to its existence that the Burmese species should possess a white eyebrow, while the Indian birds seem to fare excellently without that ornament.

Except at the nesting season, the habits of pied starlings are very like those of the other species of myna. They feed largely on the ground, over which they strut with myna-like gait—no myna would dream of losing its dignity to the extent of hopping. They feed largely on insects, but will also eat fruit. They do not, as a rule, gather together in such large companies as most kinds of starling, but in places where pied mynas exist two of them, at least, usually attach themselves to each flock of the common species.

I am inclined to think that *Sturnopastors* pair for life, but that does not prevent them from performing the antics of courtship at the nesting season. This is a fact of some importance, for if birds that are mated for life indulge every year in what we call courtship, it is obvious that the commonly accepted explanation of the meanings of the antics of birds at the breeding season is a mistaken one. The accepted interpretation of these facts is that the cocks deliberately set themselves to "kill the girls," and to this end cut mad

capers and perform the other absurdities that characterise the amorous swain. I incline to the view that,
although birds select their mates, the songs and the
dances and the displays of the males are not so much
attempts to captivate the females as expressions of the
superabundant energy that literally bubbles over at
the breeding season. A ruff when courting is obviously
as mad as the proverbial hatter : he will display
all his splendours as readily to a stone as to a reeve.
At the season of love-making one frequently sees one
pied myna—presumably a cock—puff out his feathers
and inflate his throat, and then strut after another bird
just as the little brown dove (*Turtur cambayensis*) does
when on matrimony intent. At another phase of the
courtship of the pied mynas two birds will sit, side by
side, on a perch and bow and sing to one another just
as king crows (*Dicrurus ater*) do.

Most species of myna breed early in the hot weather,
but the pied mynas invariably wait until the first rain
has fallen before they set about the work of nest-
building. Colonel Cunningham suggests that the reason
for this peculiarity of the pied starling is that, as it
does not nestle in a hole but builds in a tree, it requires
the green leaves coaxed forth by the rain as a protection to its nest. If the nursery of the pied myna were
a neatly constructed cup, something might be said
for this idea, but no amount of foliage could hide from
view the huge mass of straw and rubbish that does
duty for the nest of this species. Pied mynas rely on
their pugnacity, and not on concealment, for the protection of the nest. A list of the various materials

utilised by nesting *Sturnopastors* would include almost every inanimate object which is both portable and pliable; feathers, rags, twigs, moss, grass, leaves, paper, bits of string, rope and cotton, hay and portions of skin cast off by snakes, are the materials most commonly employed. The nest is not, as a rule, placed very high up. Sometimes it is situated in quite a low tree. Once when visiting the gaol at Gonda in the rains I observed a pair of pied mynas nesting in a solitary tree which grew in one of the courtyards inside the gaol walls. Like most of its kind, the pied starling displays little fear of man. The eggs of this species are a beautiful pale blue. Blue is the hue of the eggs of all species of myna. The fact that, notwithstanding its open nest, the eggs of the pied myna do not differ in colour from those of its brethren which nestle in holes, is one of the facts that the field naturalist comes across daily which demonstrate how hopelessly wrong is the Wallaceian view of the meaning of the colours of birds' eggs.

XXXVI

A BIRD OF THE OPEN PLAIN

IT is the fashion for modern writers of books on ornithology to divide birds according to the localities they frequent, into birds of the garden, birds of the wood, birds of the meadow, birds of the waterside, etc. The chief drawback to such a system of classification, which is intended to simplify identification, is that most birds decline to limit themselves to any particular locality.

There are, however, some species which are so constant in their habits as to render it possible to lay down the law regarding them and to assert with confidence where they will be found. Of such are the finch-larks. I have never seen a finch-lark anywhere but on an open uncultivated plain or in fields that happen to be devoid of crops.

Any person living in India may be tolerably certain of making the acquaintance of the ashy-crowned finch-lark (*Pyrrhulauda grisea*) by repairing to the nearest open space outside municipal limits.

The finch-lark is a dumpy, short-tailed bird, considerably smaller than a sparrow. Having no bright colours in its plumage, it is not much to look at, but it makes up by its powers of flight for that which it lacks in form and colour.

The finch-larks found in India fall into two genera, each of which is composed of two species.

The commonest species is that mentioned above—the ashy-crowned or, as Jerdon calls it, the black-bellied finch-lark.

In the genus *Pyrrhulauda* the sexes differ much in appearance, while in the allied genus, *Ammomanes*, the cock is indistinguishable from the hen.

As the habits of these two genera are alike in all respects, they afford an instance of the futility of attempting, as some do, to account for the phenomenon of sexual dimorphism by alleging that the habits of the dimorphic species differ from those of the mono-morphic species. When species A lives in the same locality as species B, nests at the same season, builds the same kind of nest, and when both feed and fly in the same manner, it should be obvious to every person not obsessed by a pet theory that natural selection cannot have had much to do with the fact that, whereas in species A the sexes are alike, in species B they differ. But, as we shall see, finch-larks would almost seem to have been created expressly to upset present-day zoological theories.

Well might one say to the indoor naturalist, who sits in his chair and theorises, "Go to the finch-lark, thou sluggard, consider her ways, and be wise."

The cock *Pyrrhulauda grisea* is an ash-coloured bird with a short brown tail, and very dark brown, practically black, chin, breast, and abdomen. The

cheeks are whitish, as are the sides of the body ; but
these are separated by a black bar, so that the bird
has stamped on its breast a black cross. There is also
a black or very dark brown bar that runs from the
chin through the eye. The hen is an earthy-brown
bird, the plumage being tinged with grey above and
reddish below. There is nothing peculiar in her
colouring. But for her size, she might pass for a hen
sparrow. The colouring of the cock, however, is very
remarkable. Almost every bird in existence, which is
not uniformly coloured, is of a much lighter hue below
than above. In the cock finch-lark this relation is
reversed. I cannot call to mind any other Indian bird,
unless it be the cock brown-backed robin (*Thamnobia
cambaiensis*), in which this phenomenon occurs. Moreover, the arrangement of colour—dark above and pale
below—is not confined to birds,but runs through nearly
the whole of the animal kingdom. So much so that
Mr. Thayer asserts that the phenomenon is a striking
example of protective colouration. The fact that a
bird or mammal is darker in hue above than below
renders it less conspicuous than it would be were it
coloured alike all over, since the pale under parts tend
to counteract the effects of light and shade. A few
creatures, as, for example, the skunk in America, are
darker below than above. These are usually cited
as examples of warning colouration. The skunk, as
everyone knows, is able when attacked to eject a very
fœtid and blinding excretion, so that very few animals
prey upon it. Consequently, the light-coloured back
and the erect tail are supposed to act as danger signals

to its fellow-creatures. However, there are a number of nocturnal mammals, such as our Indian ratel (*Mellivora indica*), of which the fur is light-coloured above and dark below. These cannot be examples of warning colouration. The same must be said of the inoffensive little finch-lark, with its dark under parts.

The fact that there exist so few creatures of which the under parts are of darker hue than the upper parts must, I think, be attributed to two causes. The first is that few species ever vary in that manner; the tendency is all the other way. The second is that such rare variations, when they do occur, are in most cases not conducive to the welfare of the individual, since they tend to make it conspicuous to its foes or its quarry. In certain cases, however, as in that of creatures like the shunk, which are not preyed upon, or that of nocturnal animals, the possession of dark under parts does not affect the chances of the possessor in the struggle for existence. So this variation has not been eliminated by natural selection. This, I believe, is the case with the finch-lark. The bird has very short legs, so that when it is on the ground its black under parts are scarcely visible even to a human being walking on the ground, and certainly would not be seen by a bird of prey flying overhead. My experience is that the cock finch-lark is not more conspicuous than the hen. Both, when they alight on a ploughed field, are lost to human sight until they move.

I believe finch-larks feed exclusively on the ground. I have not seen one perch in a tree. What they live

upon I do not know. The books do not tell us, and I
have never had the heart to shoot one of these small
birds in order to find out. But whatever their food
consists of, the search for it leaves finch-larks plenty
of leisure, much of which they spend after the manner
of the skylark clan. Suddenly one of these birds will
jump into the air, and rise almost perpendicularly by
vigorous flappings of its powerful little wings. Having
reached an altitude of from twenty to forty feet, its
habit is to close its pinions and drop, head foremost,
like a stone. Just before it reaches the ground, it
checks its flight and again soars upwards. Often while
disporting themselves in the air these birds display
strange antics, twisting and turning about much as the
common fly does. After amusing themselves for some
time in this manner, the pair will take to their wings
in real earnest, and fly off to a spot a quarter of a mile
or more away, and there drop to the ground and begin
feeding.

Finch-larks, like skylarks, nest on the ground.
According to Hume, they have two broods, one in
February or March, and the other in July or August.
The nest, which consists of a small pad of dried grass
and fibres, is usually placed in some depression on the
ground ; a hoof-print is considered an especially suitable site. As the bird sits very close, the nest is not
easy to find. But when flushed the hen generally
flies straight off the nest without first running along
the ground ; thus, if the spot from which the bird gets
up be carefully marked, the nest ought to be found
without much difficulty.

Finch-larks sometimes entertain queer notions as to what constitutes a desirable nesting site. At Futteh-garh Mr. A. Anderson once found a nest "in the centre of a lump of cow-dung, which must have been quite fresh when some cow or bullock ' put its foot in it.' " " As the foot-print," writes Mr. Anderson, " had not gone right through to the ground, I was enabled to remove the lump of dung without in any way hurting the nest. White ants had left their marks all over the dry dung, so that detection was almost impossible : it was altogether the most artfully concealed nest I have ever seen." Scarcely less objectionable, from the human point of view, was the site of the finch-lark's nest found at Etawah by Hume, namely, on the railway line, amongst the ballast between the rails. " When we think," says Hume, " of the terrible heat glowing from the bottom of the engine, the perpetual dusting out of red-hot cinders, it seems marvellous how the bird could have maintained her situation." Verily, there is no accounting for taste ! Two eggs are laid, which are like miniature lark's eggs.

The other species of finch-lark found in South India is *Ammomanes phœnicura*, the rufous-tailed finch-lark. This, as its name indicates, has a reddish tail. The rest of the plumage is brown. The sexes are alike. Its habits are those of the ashy-crowned species. I have not observed it in the vicinity of Madras.

XXXVII

BIRDS IN THE COTTON TREE

LACK of green grass and the paucity of wild flowers are the chief of the causes which render the scenery of the plains of India so unlike that of the British Isles. India, not being blessed with frequent showers, the *sine qua non* of flower-decked, verdant meadows, has to be content with a xerophilous flora. But there is in this country some compensation for the lack of flowers of the field in the shape of flowering shrubs and trees. Among the most conspicuous of these is the cotton tree (*Bombax malabaricum*). This tree is not an evergreen. It loses its leaves in winter, and before the new foliage appears the flowers burst forth—these may be bright red or golden yellow. As they are larger than a man's fist, and appear while the branches are yet bare, a cotton tree in flower is a very conspicuous and beautiful object. But it is of the feathered folk that visit this tree that I would write, not of the splendour of its blossom. Even before the March sun has risen and commenced to dispel the pleasant coolness of the night the cotton tree is the scene of riot and revelry. Throughout the morning hours, as the burning sun mounts higher and higher in the hard blue sky, the revelry continues. It may, perhaps, cease for a time during the first two hours after noon, when the wind blows like a blast from a titanic furnace. But it soon recommences, and not until the sun has set in a dusty haze, and the harsh clamours of the spotted owlets (*Athene brama*) are heard, does the noisy assembly of brawlers leave the tree in peace.

The cause of all the revelry is this. The nectar which the great red flowers secrete is to certain birds what absinthe is to some Frenchmen. First and foremost, amongst the votaries of the silk-cotton tree are the rose-coloured starlings (*Pastor roseus*). During the winter months these birds are not a conspicuous feature of the India avifauna, for they do not then go about in great flocks. But from the time the cotton tree is in blossom until the grain crops are cut, the rosy starlings vie with the crows in obtruding themselves upon the notice of human beings in Northern India. You cannot ride far in the month of March without hearing these birds. Their clamour is truly starling-like ; they produce that curious harsh sibilant sound which is so easy to recognise, but so difficult to describe, that noise which Edmund Selous calls a murmuration, and which the countryfolk at home term a " charm," meaning, as Richard Jefferies expresses it, " a noise made up of innumerable lesser sounds, each interfering with the other."

Look in the direction whence the sound issues and a blaze of scarlet will meet the eye ; it is amid this that the rosy starlings are calling, for where the silk-cotton tree is in bloom there are these birds certain to be.

P

Approach the tree and look carefully into it ; you will see it thronged with birds, mainly rosy starlings. Conspicuously arrayed though these birds are, it is not easy, unless they move, to distinguish them among the red petals and dark calyces. *Pastors* that are not dipping their heads into the red shuttlecock-like flowers are all either scolding one another or making a joyful noise. They move about so excitedly and jostle one another so rudely as to give you the impression that they are somewhat the worse for liquor. This may not be so. It may be the natural behaviour of the rosy starlings, for they are always noisy and pugnacious. But they seem to be exceptionally so when in the silk-cotton tree. So eagerly do they plunge their beaks into the cup-like flowers, that these latter are frequently knocked off the stalk in the process. This is especially the case with those flowers that have begun to fade. The floral envelopes and the stamens of such are easily detached from the ovary.

The rose-coloured starlings are by no means the only members of the clan which drink deeply of the nectar provided by this hospitable tree. Among the mob of brawlers are to be seen the common, the bank, and the Brahminy mynas, but there is this difference between these latter and their rose-coloured brethren ; the former are only occasional visitors to the tree. They are moderate drinkers ; they visit the public-house perhaps but once in the day, stay there a short time, and then go about their business. The rosy starlings carouse throughout the hours of daylight.

Another *habitué* of the silk-cotton tree is the Indian

tree-pie (*Dendrocitta rufa*), the nearest approach we have to the magpie in the plains of India. His long tail and general shape at once stamp him as a magpie, but his colouring is, of course, very different ; in place of a simple garment of black and white he exhibits black, chestnut-brown, silver, white, and yellow in his coat of many hues. You are not likely to see a crowd of tree-pies among the red blossoms, for the simple reason that the species is not gregarious ; but in all localities where tree-pies exist you may be tolerably certain of seeing at least one of these birds at every flowering cotton tree. Tree-pies, be it noted, although widely spread in India, are apparently very capriciously distributed. For some reason which I have not been able to fathom they occur in the neighbourhood of neither Madras nor Bombay.

Needless to say, the crows join in the drinking bout. The corvi rarely wander far from the path of the transgressor. Fortunately for the starlings, the crows are not passionately fond of the secretion of the Bombax flowers. Did these last exercise so great an attraction for the crows as they do for starlings, the smaller birds would be crowded out by their larger rivals, and the Bombax tree would be black with squawking corvi. The crow drinks the nectar of the cotton tree as a man drinks liqueurs ; the result is that rarely more than two or three crows are to be seen among the scores of starlings and mynas. The flowing bowl seems to have greater attractions for the corby (*Corvus macrorhynchus*) than for the house crow (*C. splendens*) ; but there is a reason which prevents the too frequent visiting of the silk-cotton tree by the corbies, namely, that it comes into flower in March, which happens to be the nesting season of those birds.

The above seven species are, so far as my observation goes, the only birds that make a habit of drinking at the blossom of the cotton tree. It would thus appear that the nectar has a very pronounced taste, and that, in consequence, birds either like it intensely or positively dislike it.

" Eha," I am aware, states that many other birds frequent the cotton tree, for the sake of its good cheer, " the king crow, and even the temperate bulbul and demure coppersmith, and many another, and, here and there, a palm squirrel, taking his drink with the rest like a foreigner." But did not " Eha " mistake the purpose for which these creatures visit the silk-cotton tree ? A bird may be present without taking part in the revelry. The other day I was watching all the fun at one of these trees when suddenly a little coppersmith (*Xantholæma hæmatocephala*) came and perched on one of the bare spiny branches. He sat there motionless, as out of place as a Quaker would among a mob of bookmakers. Suddenly a rosy starling hustled him off his perch. But the coppersmith did not fly away ; he merely hopped on to another branch, and then suddenly performed the vanishing trick. Had I not been watching him very closely I could almost have persuaded myself that he had melted into thin air. As it was, I saw him dive into a round opening— scarcely the size of a rupee—about two inches from the broken end of a dead branch, not as thick as a woman's wrist, at the very summit of the tree. The circular opening in question had been neatly cut by the coppersmith and its mate, and led to a hollow in which three white eggs were doubtless lying. These and not the nectar-bearing flowers were the attraction for the coppersmith.

XXXVIII

UGLY DUCKLINGS

SOME people invariably look untidy. They seem to be nature's misfits. All the skill of the tailor, all the art of the milliner, can make them nothing else. No matter how well-cut their garments be, these always hang about them in a ridiculous, uncouth manner. If the individual be a man, the upper part of his collar seems to exercise an irresistible attraction for his tie ; if a woman, she presents an unfinished appearance about the waist, as often as not displaying an ugly hiatus in that region. Similar creatures are to be found among the beasts of the field and the birds of the air. There exist not a few feathered things whose plumage usually looks as though a thorough spring-cleaning, followed by a " wash and brush-up," would do it a world of good. Chief among these are our well-known friends the babbler thrushes, alias the seven sisters, or seven brothers, as some will have it.

Like most human beings who are careless of their personal appearance, these birds possess many good qualities. First and foremost of these is the love which they show one to another. They are brotherly affection personified. Except for a very rare squabble over a tempting piece of food, the harmony of the brotherhood is never broken. What more striking testimony to this admirable quality can be offered than the popular designation of the bird ? It is always one of seven ; there is no word whereby the man in the street may express an individual alone without his comrades. Nor, indeed, does he require such a term, for it is impossible to think of the bird otherwise than as one of a company. Has anyone ever seen brother Number One, or brother Number Two, or brother any other number alone ? I trow not. These birds invariably hunt in little societies ; usually eight or ten elect to fight the battle of life shoulder to shoulder, and a very good fight they appear to make of it, if we may judge by their wide distribution and contented faces.

While upon the subject of the bird's name it is as well to have the usual hit at the ornithologist. Just as the popular name is appropriate, so is the scientific one ridiculous. *Crateropus canorus* is a strange name for a bird whose note is a cross between the creak of a door with a rusty hinge and the squeak of a cartwheel of which the axle needs oiling. Nature, by way of compensation, often endows a sombre-plumaged bird with a sweet voice, and keeps down the pride of a gorgeous fowl by ordaining that its voice shall be a hoarse croak. To the seven brothers, however, the wise dame has given two wooden spoons. Their raucous voice is in keeping with their dull plumage. When the honest little company are merely whispering

sweet nothings one to another, the stranger un-
acquainted with their habits is apt to think that they
are angrily squabbling, and that bloodshed must
inevitably follow. Such is the voice of the bird yclept
" canorus " by the ornithologist.

Linnæus appears to have given this species this name
under the impression that it was the Indian equivalent
of our English thrush, that it sat in mango trees and
warbled most sweetly.

Hodgson made a gallant attempt to give the species
the more appropriate name " terricollor," but he
laboured in vain. The tyranny of the priority rule
proved too much for him.

Ornithological public opinion has decreed that as
regards the specific names of birds the race is to the
swift : the first name hurled at a bird, no matter
how inappropriate, is to be retained. This rule was
made in the hope of introducing some sort of order into
the chaos of ornithological terminology. But, far from
effecting this, it has called into existence a race of
ornithological pettifoggers, who spend their time in
rummaging about in libraries in the hope of discovering that some bird bears a name which was not
the first to be given it. Such a discovery means another
change in ornithological terminology. This is provocative of much unparliamentary language on the
part of the naturalist, but gives the priority-hunter
unalloyed pleasure.

Is it necessary for me to describe these misnamed
babblers ? Who is not familiar with the untidy
creature, with his dirty-looking brownish-grey plumage,

relieved by a yellow beak and a white, wicked eye ?
Who has not laughed at the drooping wings, the
ruffled feathers, and the disreputable tail of the birds ?
Yet the seven brothers lead happy, contented lives.
They have always company, and plenty to occupy
their minds. They are numbered among those who
despise not small things : no insect is too tiny, no
beetle too infinitesimal, no creeping thing too insignificant, to be eaten by these birds, so the little
company of friends hops together along the ground
from tree to tree, from shrub to shrub, searching every
nook and cranny, turning over every fallen leaf in the
most methodical way, seizing with alacrity everything
it comes across in the shape of food. During the search
for food the chattering never ceases. Now and again
the birds will take to a tree and hop about its branches,
talking louder than ever. In the early morning, before
the air has lost its first crispness, they delight to play
about the trees, flying in a crowd from one to another.
Again, in the evening, just before bedtime, they love
to gambol among the branches and jostle one another
in the most good-tempered way.

These birds have adopted the motto of the French
Republic, and they practise what they preach. Liberty,
equality, and fraternity are theirs. They form a true
republic, a successful one because of the smallness of
its numbers. What bird is so free as our seven
brothers ? They are not hedged in by the conventions
of dress. " Eha " says that they remind him of " old
Jones, who passes the day in his pyjamas." Is this not
the acme of freedom ? They squeak, croak, hop, and

fly where they list ; well might they be enrolled in the Yellow Ribbon Army, that noble band who eat what they like, drink what they like, say what they like, and do what they like.

Of the fraternity of the little society we have already spoken. Of their equality there can be no room for doubt. They have no leader. Now brother Number Two, now brother Number Five moves on first, to be followed by his comrades. They seem all to fall in with the views of whoever for the moment takes the lead. There is much to be said for this form of life. It makes the birds, who are individually weak, bold. They have often hopped about outside my tent, jumping on to the ropes, and seeking food within a couple of inches of the *chik* on the other side of which I was standing. They seem to court the company of man. It is in the compound rather than the jungle that they abound. If one of the little company be attacked by a more powerful bird, his comrades come at once to his assistance. Some naturalists declare that they will go so far as to attack a sparrow-hawk, others say they will not. Probably both are right. All men are not equally brave, nor are all babbler thrushes equally bold. Even the bravest species has to confess to a Bob Acres or two. As a matter of fact, the brotherhood is not afforded many opportunities of displaying its valour, for it is rarely attacked. Birds of prey know better than to molest social birds ; they are aware of the fact that it is difficult to elude sixteen or twenty watchful eyes, and even if this feat be accomplished there is always the fear of a stout resistance. The babbler thrushes recall the good old days of ancient Rome when all were for the State and none for a party.

The seven brothers are as indifferent to the appearance of their home as to that of their persons. The nest they construct is a rude structure, but some species of cuckoo think it quite good enough to lay eggs in.

XXXIX

BABBLER BROTHERHOODS

THE Crateropus babblers, known variously as the *Sath Bhai*, seven sisters, or dirt birds, furnish perfect examples of communal life. So highly developed are their social instincts that a solitary babbler, or even a pair, is a very unusual sight. They do not congregate in large flocks ; from six to fourteen usually constitute a brotherhood, eight, nine, or ten being, perhaps, the commonest numbers. There is no truth in the popular idea that they always go about in flocks of seven. Sir Edwin Arnold recognised this when he wrote of " the nine brown sisters chattering in the thorn."

Notwithstanding the fact that babblers are among the commonest birds in India, there is much to be discovered regarding the nature of their flocks. The *raison d'être* of these flocks is not far to seek. One has but to observe the laboured flight of a babbler to appreciate how easy a mark he is to a bird of prey. The strength of the babbler lies in his clan. Eight or ten pairs of eyes are superior to one. A party of seven sisters is not often caught napping. The incessant squeaking, and screeching, and wheezing indulged in by each member keep them all in touch with one another. Then, in time of danger, it is good to see how they combine to drive off the hawk-cuckoo (*Hierococcyx varius*) which victimises them, and which they undoubtedly mistake for a species of raptorial bird.

But their clannishness does not shelter them from all tribulation. They are the dupes of the hawk-cuckoo, and they sometimes fall victims to birds of prey. A few weeks ago I had occasion to visit a friend, who was unwell and confined to his bungalow. I found him sitting in the verandah. While greeting him I heard a great clamour of scolding babblers (*Crateropus canorus*) emanating from a neem tree hard by. I had come just too late to witness a little jungle tragedy. There was a babbler's nest containing young in that tree. A pair of rascally crows had discovered the nest, and one of them attacked it ; the babbler in charge, with splendid courage, went out to meet his big antagonist, who promptly turned tail and fled, pursued by the screeching babbler. This left the nest open to the other crow, who carried off a young bird. When I arrived, the victims of the outrage were swearing as only babblers and bargees can, and making feints at the crows.

It is thus obvious why these clubs, or brotherhoods, have been formed, but we are almost altogether in the dark as to how they are formed, as to their nature and constitution. We do not even know what it is that keeps them apparently so constant in size. It is even a disputed point whether these little companies persist throughout the year, or disband at the nesting season.

As to the nature of the companies, Colonel Cunningham maintains that they are family parties. This view is, however, untenable, unless we assume that the seven sisters are polygamists or polyandrists, because three or four is the normal number of eggs laid, so that if each little gathering were a family party, it should consist of not more than six members. The flocks are too large to be made up of mother, father, and children, and usually too small to be two such families.

There is at present living in the compound of the Allahabad Club a company consisting of, I think, eight babblers. Seven are adults, and one is quite a child. This last goes about with its elders, every now and again flapping its wings, opening wide its yellow mouth, and calling for food. A day or two ago it took up a position within a few feet of my door, so that I was able to watch it closely through the *chik*. I saw one of the company come up with a grub in its bill, which it, with due ceremony, put into the young bird's " yellow lane." Having fed the youngster, it began rummaging about in the grass near by. Shortly afterwards a second babbler came up to the young one, bringing a caterpillar. This particular individual carried his (or her, for I don't pretend to be able to sex a babbler at sight) tail askew. That organ protruded from under the left wing, instead of projecting between the wings, as is usual with tails—babblers, like actors and artists, affect a careless style of dress. Having delivered up its caterpillar to the clamorous youngster, it hopped away. I kept my eye carefully upon both it and the bird I had first seen bring food. In a few seconds a third babbler came up and presented a caterpillar to the baby brown sister. Now, I submit that this can only mean that babblers are not monogamous, or that they nest in common sometimes, or, so close are the ties that bind the members of the little company that each feeds both his own offspring and those of his brethren. Personally, I am inclined to think that babblers are monogamous. That the same nest is sometimes used by more than one pair seems to be established by the fact that there are cases on record of nests containing as many as eight eggs, or young ones. This, however, is not a usual occurrence, and it is my belief that the members of the club are so greatly attached to one another that they look upon each infant as common property. Hume quotes Mr. A. Anderson as saying : " During the months of September and October I have observed several babblers in the act of feeding one young *Hierococcyx varius* (the brain-fever bird or hawk-cuckoo, which, as we have seen, is parasitic on babblers) following the bird from tree to tree, and being most assiduous in their attentions to the young interloper." This observation, I submit, supports the view that each member of the flock takes a personal interest in the offspring of other members, even though it be spurious !

Thus we may take it that these gatherings are not family parties, but rather of the nature of clubs. The question, then, arises : What determines the membership of these clubs ? At present our knowledge of the ways of these common birds is not sufficient to enable us to frame a satisfactory reply. It is even an open question whether or not these clubs break up at the breeding season, or whether the nesting birds still continue to seek food in company. Colonel Cunningham declares that during April and May babblers " cease to go about in parties, and pairs of them are everywhere busily occupied in nesting." Jerdon, on the other hand, states that the parties persist throughout the breeding season. I feel sure that Jerdon is right. No matter where one is stationed, parties of babblers are to be seen at all seasons of the year. From this, of course, it does not necessarily follow that the nesting birds do not forsake their brethren, at any rate for a time. It is probable, nay certain, that all the members of a flock do not pair and nest simultaneously. The breeding season extends at least from March to July. But the fact that there is quite a baby bird in the babbler brotherhood that dwells in the compound of the Allahabad Club seems to indicate that the nesting birds continue to find their food in company. There is no reason why they should not, for babblers neither migrate nor wander far afield.

But the question arises: What happens to the young birds when they are grown up? If they attached themselves to the existing flocks, these would tend to increase in size, and sometimes, at any rate, we should see an enormous assembly. So far as one's casual observation goes, the flocks keep constant in number throughout the year. It is, of course, quite possible that casual observation leads one astray in this case. Any person interested in the subject, who has a more or less fixed abode, would do some service to orni-

thology if he would make a point of looking out for
babbler clubs, and endeavouring to count the members
of each, and keep a record of the results, with the
date of each census. I am aware that it is not easy to
count accurately a babbler club, for its members are
always on the move, and odd birds are apt to pop out
of unexpected places. But even rough figures, if they
extended to a number of flocks, would, being all liable
to the same error, prove fairly accurate as regards
averages. Such observations, if they were to extend
over a year, might lead to some interesting results.
They would almost certainly show a reduction of
numbers during the summer months, when nesting
operations were in progress, but would this be followed
by a considerable rise later in the year ? If so, it
would seem to indicate that some, at any rate, of the
young ones attached themselves permanently to the
flock in which they were born.

A somewhat more elaborate experiment which might
yield interesting results would be to trap a whole
" school " of babblers ; they might be captured while
asleep. After a piece of coloured material had been
tied round the leg of each, every bird being decorated
by a different colour, the irate sisters would be restored
to liberty. Then it might be possible to follow the
fortunes of each separate bird, and learn whether a
given flock is always made up of the same individuals,
whether they practise exogamy or favour endogamy,
and a hundred and one other interesting facts relating
to the *vie intime* of the brown sisters. I use the
word " might " advisedly. For alas ! bitter experience

Q

has taught me that, more often than not, the most cunningly devised ornithological experiments yield no definite results. It is quite possible that the club of babblers thus captured and decorated with gay colours might flee from the neighbourhood in wrath and terror. The birds would not understand the why and the wherefore of the proceeding, and might, perhaps, think that you were going to make a practice of catching them every night and tying things round their limbs. A bird whose leg has been pulled once is apt to be twice shy.

XL

THE MAD BABBLER

THE seven sisters (*Crateropus canorus*), which occur in every garden in India, are veritable punchinellos, so much so that schoolboys in the Punjab always call them "mad birds." But nature is not content with having produced these. So readily does the babbler clan lend itself to the humoresque, that from it has been evolved the large grey babbler (*Argya malcomi*), a species even more comic than the noisy sisterhood. This is the *Verri chinda*, the mad babbler of the Telugu-speaking people. Pull the tail out of one of the seven sisters, and insert in its place another, half as long again, with the outer feathers of conspicuously lighter hue than the median ones, then brush up the plumage of the converted sister, and you will have effected a transmutation of species, turned a jungle babbler into a large grey one. This latter species has a wide range, but is capricious in its distribution. It does not, I believe, occur in the neighbourhood of the city of Madras, but is abundant in some parts of South India. The habits of this species seem to vary with the locality. In the south it appears to shun the madding

227

crowd ; in the north it frequents gardens and loves to
disport itself in the middle of the road, and is in no
hurry to get out of the way of the pedestrian or the
cyclist.   Probably many a large babbler has, owing
to its tameness, succumbed to the motor-car.   Bold
spirits, such as the little striped squirrel, which take
a positive delight in experiencing a series of hair-
breadth escapes, suffer considerably when a new and
speedier conveyance is introduced into a locality.
They have learned by experience how close to the
inch they may with safety allow the ordinary vehicle
to approach before they skedaddle, and it takes time
for them to discover that with a speedier vehicle a
larger margin must be allowed.   The little Indian
squirrel has not yet learned to gauge the pace of the
motor-car.   Recently I counted five of their corpses on
the road between Agra and Fatehpur Sikri, which is
much frequented by motor-cars.

The *Sath Bhai* are usually accounted noisy birds, but
they are taciturn in comparison with their long-tailed
cousins.   From dewy morn till dusty eve the large
grey babblers vie with the crows in their vocal efforts.
The crows score at the beginning of the day, for they
are the first to awake, or, at any rate, to begin calling.
The king crow (*Dicrurus ater*) is usually said to be the
first bird to herald the cheerful dawn.   This is not
always so ; the voice of *Corvus splendens* sometimes precedes that of the king crow.   But ere the sun has
shown his face the grey babblers join vociferously in
the chorus that fills the welkin.   And how shall I
describe the notes of these light-headed birds so as

to convey an adequate idea of them to those who have not heard with their own ears ? I ought to be able to do so, for Allahabad, where I am now stationed, is the head-quarters of the clan of large grey babblers. *Argya malcomi* are to that city what the Macphersons are to Inverness-shire. You cannot avoid them. The sound of their voices is never out of my ears during the hours of daylight. Some of them are shouting at me even now. Yet words to describe what I hear fail me. The only instrument made by man that can rival the call of the mad babbler is the "rattle" used at our English Universities, or at any rate at Cambridge, to encourage the oarsmen in the Lent or May races. It is the delight of two of these birds each to take up a position at the summit of a tree and for one to commence calling. He bellows till his breath runs short; then his neighbour takes up the refrain—I mean, hullabaloo—and, ere number two has ceased, number one, having recovered breath, chimes in. In addition to this rattle-like call the grey babblers emit a more mellow note, which is well described by Jerdon as " Quey, quey, quey, quo, quo," pronounced gutturally. Occasionally one of these extraordinary birds bursts out into a volley of excited squeaks, like the voice of Punch as rendered by the showman at the seaside. This I take to be a cry of alarm. The bird while uttering it careers about madly among the foliage of a tree, hopping from bough to bough with great dexterity.

Mad babblers go about, like the seven sisters, in flocks of ten or twelve, and feed largely on the ground.

Their mode of progression when not on the wing is by a series of hops. Their movements are very like those of a thrush on an English lawn—a dash forward for about a yard, followed by an abrupt halt. They seem to subsist chiefly on insects, but grain does not come amiss to them. In places where they abound, several of them are usually to be seen in every field of large millet, each perched at the summit of a stalk eagerly devouring the ripening grain. When thus occupied they sometimes forget to call. They are birds of peculiarly feeble flight. Their tail is long and their wings are somewhat sketchy, and the result is that they have to flutter these latter frantically in order to fly at all. But for the fact that they always keep together in flocks, even at the nesting season, they would fall easy victims to birds of prey. Thanks to their clannishness and pluck, they appear to be tolerably immune from attack. Jerdon says : " If the Shikra sparrow-hawk be thrown at them, they defend each other with great courage, mobbing the hawk and endeavouring to release the one she has seized." Only yesterday I saw a party of about a dozen large grey babblers attack and drive away a couple of black crows (*Corvus macrorhynchus*) from a position which the latter had taken up on the ground. The babblers advanced slowly in a serried mass, while the corbies remained motionless watching them. When the front rank of the babbler *posse* had advanced to within a foot of the crows a halt was called, and the adversaries contemplated one another in silence for a few seconds. Then one of the babblers made a lunge at the corby, which

caused it to take to its wings. Immediately afterwards
the other crow was similarly driven away. While the
babblers were still celebrating their bloodless victory
with a joyful noise, a tree-pie (*Dendrocitta rufa*) came
and squatted on the ground near them, evidently
spoiling for a fight. The babblers advanced against
him, this time in open order. On their approach the
pie lunged at a babbler and caused it to retire. But
immediately another babbler made a feint at the tree-
pie, and things were becoming exciting when something scared away the combatants.

Argya malcomi constructs a nest of the typical
babbler type ; that is to say, a somewhat loosely
woven cup, which is placed, usually not very high
above the ground, in a tree or bush. Nests are most
likely to be found in the rains. The eggs are a beautiful
rich blue—the hue of those of our familiar English
hedge-sparrow (*Accentor modularis*)—which is so characteristic of babblers.

Like all of us, this happy-go-lucky babbler has its
trials and troubles. It is victimised by that handsome, noisy ruffian, the pied crested cuckoo (*Coccystis
jacobinus*), which deposits in the nest an egg, which is
a first-class counterfeit of that of the babbler. The
feckless babblers sit upon the strange egg until it gives
forth its living contents. The presence of the spurious
child does not greatly perturb the babblers. As we
have seen, the flock does not break up even at the
nesting season. Under such circumstances the whole
flock probably takes part in administering to the young
cuckoo the wherewithal to fill the inner bird, so that on

the principle " many hands make light work " the extra mouth to feed is scarcely noticed. But is it an extra mouth? Does the young pied cuckoo eject its foster-brethren, or do the parents turn out the legitimate eggs?

XLI

THE YELLOW-EYED BABBLER

THE babbler community embraces a most heterogeneous collection of birds. Every Asiatic fowl which does not seem to belong to any other family is promptly relegated to the Crateropodidæ. Thus it comes to pass that such dissimilar creatures as the laughing thrushes and the seven sisters find themselves classed together. Now, taken as a whole, the babbler class is characterised neither by beauty nor melodiousness. The best-known members are the widely distributed seven sisters, which in many respects are very like those human babblers who style themselves Labour Members of Parliament. They are untidy in appearance and exceedingly noisy ; their voices are uncouth, and they never tire of hearing themselves shout. They are apt to meddle with affairs that do not concern them. Of course the *Sath Bhai* have their good points ; so, I suppose, have Labour M.P.'s—at any rate when they are in their natural habitat. When they come to India and then try to wield the pen—but it is not of human babblers that I wish to write, nor of the plainly attired, noisy, avian babblers, for have not the seven sisters had a double innings already ? Even as some Labour

Members of Parliament wear frock-coats and top hats,
so are there some well-dressed members of the babbler
clan. The yellow-eyed babblers belong to this class ;
and the most widely distributed of these—*Pyctorhis
sinensis*—is the subject of the present discourse. This
bird is, according to Oates, found in every portion of
the Indian Empire up to a height of 5000 feet. As a
matter of fact I have not seen it in or near the city of
Madras, but that, perhaps, was not the fault of the
bird, because we have Jerdon's testimony that he saw
it in every part of South India.

The yellow-eyed babbler is a sprightly little creature
not much larger than a sparrow. Its upper plumage
is a rich reddish brown, changing to cinnamon on some
of the quill feathers. The chin, throat, cheeks, and
breast are as white as snow. The conspicuous orange-
yellow eye is set off by a small white eyebrow. The
abdomen is cream-coloured. The bill is black and the
legs a curious shade of dull yellow. The tail is 3½ inches
long, at least the median feathers thereof are ; the
outer ones are barely two inches in length. This gradation in the size of the caudal feathers is, of course,
visible only when the tail is spread during flight. The
yellow-eyed babblers that inhabit Ceylon differ from
those of the mainland in some unimportant details ;
hence systematists, with their usual aptitude for
species-making, call the former *Pyctorhis nasalis* to
distinguish them. In many parts of India the yellow-
eyed babbler is quite a common bird. It is especially
addicted to tall grass and hedgerows, and will occasionally enter a garden that is well provided with

bushes. It is not so clannish as most of its brethren ; sometimes a small party of six or seven feed in company, but more often only solitary birds or pairs are seen. They hop about in and out of small bushes or on the ground, industriously seeking out the small beetles and other insects on which they prey. Every now and then one of these sprightly birds permits itself a little relaxation in the shape of a sweet melody, which it composes and pours forth from the summit of a convenient bush. Its more usual note is described by Jerdon as "a loud sibilant whistle" ; it also utters a variety of chattering sounds, which proclaim it a true babbler.

For an Indian bird it is shy ; if it sees that it is being watched it quickly disappears into cover.

The nest of this species is a veritable work of art. Its usual form is that of an inverted cone, composed of dried grass, fibres, or other suitable material very compactly and neatly woven, the whole being plastered over exteriorly with cobweb, which, as I have said before, is the cement generally used by bird artisans. The well-built little nursery is sometimes wedged into a forked branch of a tree ; more often it will be found snugly tucked away in a bush. In the Punjab the nest is very frequently found attached to the stalks of growing millet, in much the same way as a reed-warbler's nest is fastened to reeds. The babbler weaves its nest round a couple of adjacent stalks, so that these are worked into its walls. A nest which is thus supported by two stalks is in shape like the cocked hat worn by a political officer.

The eggs, which may be looked for at any time between May and September, are very beautiful. To describe them in a few words is not easy, because they exhibit great diversity in colour and markings. This is one of the hundreds of facts inconsistent with the orthodox theories of the significance of colour in organic nature that confront the field naturalist at every turn. The existence of such facts does not perturb in the least those theorists who " rule the roost " in the scientific world. Their attitude is " our word is law—if facts don't fit in with it, so much the worse for facts." As Hume points out, three main types of eggs occur, and there are many combinations of these types. Of the two types most often seen, " one has a pinkish-white ground, thickly and finely mottled and streaked over the whole surface with more or less bright and deep brick-dust red, so that the ground colour only faintly shows through here and there as a sort of pale mottling ; in the other type the ground colour is pinkish white somewhat sparingly, but boldly, blotched with irregular patches and eccentric hieroglyphic-like streaks, often bunting-like in their character, of bright blood or brick-dust red."

XLII

THE INDIAN SAND-MARTIN

THE Indian sand-martin (*Cotile sinensis*) is, I believe, the smallest of the swallow tribe. So diminutive is he that you could put him in your watch-pocket, were you so minded, without fear of damaging his plumage. His charm lies in his littleness and activity rather than in his colouring, for he belongs not to the dandies. Neat and quiet are the adjectives that describe his attire. The head, shoulders, and back are pale brown tinged with grey. The wing-feathers are dark brown. The under parts are white with a touch of grey on the chin and breast. The sexes dress alike. This description applies equally well to the sand-martin (*Cotile riparia*) that nests in sand-pits in England, for the only differences between this species, which occurs sparingly in India, and the Indian form are that the former is a little larger and possesses a dark necklace.

The feeding habits of sand-martins are those of the rest of the swallow tribe. They live on minute insects which they catch on the wing, not, after the manner of fly-catchers, by making little aerial sallies from a perch, but by careering speedily through the air during the greater part of the day and seizing every insect that they meet.

The Indian sand-martin is a species especially dear to the ornithologist because it nests in winter, when comparatively few other birds are so occupied. Speaking generally, the cold weather may be said to be the " silly season " of the bird world.

There is one drawback to India from the point of view of the ornithologist, and that is the habit of the great majority of birds of building their nests at the time when the sun shines forth pitilessly from a cloudless sky for twelve hours out of the twenty-four, burning up all vegetation and raising the temperature of the air to furnace heat. Under such conditions the pleasure of watching the birds is tempered by the physical discomfort to which the bird-watcher is put. Very pleasant, then, is it, after months of excessive heat, to awake from sleep one morning to find that the cool weather has come at last, to feel the morning air blow fresh against the cheek, and to look out on an earth enveloped in dense mist. Before one's horse is saddled, the first rays of the sun dissipate the mist with almost magic suddenness, and then one rides forth over dew-bejewelled plains of grass. If on such a morning one repairs to a sand-pit or a river bank, one is likely there to meet with a colony of sand-martins, for it is early in the cold weather that those birds begin to construct their nests, which are holes bored in sand-banks by the birds themselves.

Like the majority of very small birds, sand-martins show but little fear of human beings. Tits, white-eyes, warblers, sand-martins, etc., will come in search of food quite close up to a motionless human being. Mr. W. H. Hudson relates in his *Birds and Man* how, when one day he went into his garden and walked under the trees, there was a great commotion among the little birds overhead, who mobbed him in the manner they mob an enemy. He discovered that the reason of this strange behaviour on the part of the small birds that usually paid no attention to him, was that he was wearing a striped cloth cap, which the birds appeared to mistake for a cat. It would almost seem that there is so vast a difference in size between a tiny bird and a human being that the former fails to recognise the latter as a living object provided he keeps still. This does not imply poor eyesight on the part of birds. The minds and eyes of birds are almost invariably directed on small things. Now, a man bears to a small bird much the same relation as a horse three hundred hands high would bear to a man. As regards detail, the eyesight of birds is probably superior to that of men, for each sand-martin seems never to mistake its nest, although the entrance to it is merely one of several score of holes scattered irregularly over the face of the cliff. To the human eye these holes look all very much alike, but each must possess minute peculiarities which loom large in the eye of the sand-martin. Whether or not the above explanation is the true one, the fact remains that a human being can take up a position within a few feet of the cliff without disturbing the martins in their nest-building operations.

Some birds, when busy at their nests, work with

feverish haste, as though they were under contract
to finish them by a given date. Not so the sand-
martins. With them, the spells of work at the nest
would seem to be mere interludes between their
gambols in the air. Each bird appears to visit its nest
every few seconds, but generally it contents itself with
hovering in front of the hole for a fraction of a minute
and then dashes away. Frequently one sees a martin
perch at the aperture for a few seconds without doing
any work, and then fly off again. For every visit made
with the object of doing work, ten or twelve seem to be
made for the mere fun of the thing. Sand-martins
appear to derive the greatest pleasure from the con-
templation of the growing nursery. If the cliff be
examined carefully, its soft sandy surface will be found
to be scored in many places by marks made by the
sharp little claws of the martins as the birds alight.

A colony of nesting martins presents a very animated
appearance. The main body dash through the air to
and fro in front of the cliff, uttering their feeble twittering, but a few are always at the nest holes, either resting
or working. These latter are constantly reinforced
from those on the wing, and *vice versa*, so that there
are two streams of birds, one flying to the cliff and the
other leaving it. Suddenly the whole flock, including
both the resting and the flying birds, will, as if affected
simultaneously by a common influence, fly off *en masse*
and disappear from sight. But they are never absent
for long. At the end of two or three minutes all are
back again.

The birds utter unceasingly, when on the wing, a

twittering note, not so harsh as that of the sparrow, but sufficiently harsh to make it appear that the birds are squabbling. A certain amount of bickering does take place among the sand-martins. Every now and again a bird may be observed chasing its neighbour in a very unneighbourly manner. Occasionally two will attack one another with open beak, and fall interlocked to the ground. A prettier sight is that of a couple of martins resting side by side at the orifice of the nest hole twittering lovingly to one another. The excavation that leads to the nest is a round passage, less than three inches in diameter. After proceeding inwards and slightly upwards for about two feet, it ends in a globular cavity of larger diameter. This is the nesting chamber, and is lined with grass, fine twigs, feathers, and the like. Two or three white eggs are laid. Sand-martins probably bring up more than one brood in the year. Their nests are likely to be found in all the winter months.

Cotile sinensis is a permanent resident in India and is common in all the northern portions of the country, but is not often seen so far south as Madras. It is curious that this species should be abundant in North India and rare in the south, where insect life is so plentiful. There must be something in the climatic conditions of South India that suits neither this nor the other species of sand-martin. Precisely what this is I cannot conjecture. Birds vary greatly in their adaptability to climate. Some, such as the hoopoe, appear absolutely indifferent to heat or cold, moisture or dryness ; others, as most wagtails, shun heat.

R

The two common crows of India afford an excellent
illustration of the way in which allied species differ
in their power of adapting themselves to variation in
climate. The grey-necked species (*Corvus splendens*)
is found throughout the length and breadth of the
plains of India, but does not ascend the Himalayas
to any great height, and is, in consequence, not found
in Murree Mussoorie or Naini Tal. The corby (*C.
macrorhynchus*), on the other hand, is found in all parts
of the plains save in the Punjab, and ascends the Himalayas up to 10,000 feet or higher, and is the only
crow that occurs in most of the Himalayan hill stations.
It is thus evident that the black species is far less
sensitive to cold than the other, but why does it occur
so sparingly in the Punjab ? The connection between
climate and the distribution of birds is a fascinating
subject about which very little is known. Possibly
in the varying sensitiveness of birds to climatic conditions lies the secret of some of the phenomena of bird
migration.

XLIII

THE EDUCATION OF YOUNG BIRDS

A CERTAIN school of naturalists, in which Americans figure largely, lays great stress on the way in which parent birds and beasts educate their offspring. According to this school, a young bird is, like a human babe, born with its mind a blank, and has to be taught by its parents everything that it is necessary for a bird to know. Just as children study at various educational establishments, so do young animals attend what Mr. W. J. Long calls " the school of the woods." "After many years of watching animals in their native haunts," he writes, " I am convinced that instinct conveys a much smaller part than we have supposed ; that an animal's success or failure in the ceaseless struggle for life depends, not upon instinct, but upon the kind of training which the animal receives from its mother." In short, but for its parents, a young bird would never learn to find its food, to fly, or sing, or build a nest.

This theory appears to have met with wide acceptance, chiefly because it brings animals into line with human beings. It is but natural for us humans to put anthropomorphic interpretations on the actions of animals. Careless observation seems to justify us in so doing. While not denying that birds do spend much time and labour in teaching their young, I am of opinion that the lessons taught by them are comparatively unimportant, that their teachings are merely supplementary to the instinct, the inherited education, which is latent in young birds at birth, and displays itself as they increase in size, just as intelligence develops in growing human beings.

By the mere observation of birds and beasts in their natural state it is not easy to ascertain how far the progress made by young ones is the growth of their inborn instincts, and how far it is the result of parental instruction.

It is the failure to appreciate the magnitude of this difficulty that vitiates the teachings of Mr. Long and the school to which he belongs. We can gauge the value of the pedagogic efforts of parent animals only by actual experiment, by removing young birds from parental influence and noticing how far that which we may term their education progresses in the absence of the mother and father.

The first and foremost of the things which a young bird must know is how to find its food. This is an accomplishment which it speedily acquires without any teaching. Young ducklings hatched under a barndoor hen take to the water of their own accord, and soon discover how to use their sieve-like bills.

I read some time ago a most interesting account

of two young American ospreys, which Mr. E. H. Baynes took from the nest at an early age. Having secured them, he placed them in an artificial nest which he had made for them. The parents did not succeed in finding them out, the young birds had therefore to face the struggle for existence without a mentor. " For several days," writes Mr. Baynes, " they spent most of their time lying still, with necks extended and heads prone on the floor of the nest." At this stage they were, of course, unable to fly. It was not until they were five or six weeks old that the young ospreys entrusted themselves to their wings, and at the first attempt they, or rather one of them, performed an unbroken flight of several miles ! After they had learned to use their wings, the ospreys were allowed full liberty, nevertheless they continued to remain in the neighbourhood of Mr. Baynes's house, and became quite domesticated. When taken away, they returned like homing pigeons. Even as they had made the discovery that they could fly, so did they, one day, find out that they could catch fish. Mr. Baynes thus describes the earliest attempt of one of the young birds : " His tactics were similar to those employed by old and experienced ospreys, but the execution was clumsy. After sailing over the pond for a few minutes, he evidently caught sight of a fish, for he paused, flapped his wings to steady himself, and then dropped into the water. But it was the attempt of a tyro, and of course the fish escaped. The hawk disappeared, and when he came to the surface he struggled vainly to rise from the water. Then he seemed to give it up."

At this, Mr. Baynes was about to jump into the water in order to rescue him ; however, " the next moment he made a mighty effort, arose dripping wet, and flew to his old roost on the chimney, where he flapped his wings and spread them out to dry in the sun." Far from being deterred by this experience, he repeated the operation, and ere long became an expert fisher.

According to the school to which Mr. Long belongs, young birds learn their song from their parents, just as young children learn how to talk. In the words of Barrington, " Notes in birds are no more innate than language is in man, but depend entirely upon the master under which they are bred, as far as their organs will enable them to imitate the sounds which they have frequent opportunities of hearing."

Similarly Michelet writes : " Nothing is more complex than the education of certain singing birds. The perseverance of the father, the docility of the young, are worthy of all admiration." There can be no doubt that young birds are very imitative. The young of the koel—an Indian parasitic cuckoo—make ludicrous attempts to caw in imitation of the notes of their corvine foster-parents ; but later, when the spring comes, they pour forth the very different notes of their species. In the same way the young of the common cuckoo, no matter by what species they are reared, all cry " cuckoo " when they come of age. Ducklings, pheasants, and partridges, hatched under the domestic hen, and fowls reared by turkeys, have the calls peculiar to their species. It may, of course, be urged that these learn their cries from others of their own kind. Here again, then, actual experiment is necessary to determine which view is correct. Such experiments were performed by Mr. John Blackwall as long ago as 1823. He writes :—

" I placed the eggs of a redbreast in the nest of a chaffinch, and removed the eggs of the chaffinch to that of the redbreast, conceiving that, if I was fortunate in rearing the young, I should, by this exchange, ensure an unexceptional experiment, the result of which must be deemed perfectly conclusive by all parties. In process of time these eggs were hatched, and I had the satisfaction to find that the young birds had their appropriate chirps.

" When ten days old they were taken from their nests, and were brought up by hand, immediately under my own inspection, especial care being taken to remove them to a distance from whatever was likely to influence their notes. At this period an unfortunate circumstance, which it is needless to relate, destroyed all these birds except two (a fine cock redbreast and a hen chaffinch), which, at the expiration of twenty-one days from the time they were hatched, commenced the calls peculiar to their species. This was an important point gained, as it evidently proved that the calls of birds, at least, are instinctive, and that, at this early age, ten days are not sufficient to enable nestlings to acquire even the calls of those under which they are bred. . . . Shortly after, the redbreast began to record (i.e. to attempt to sing), but in so low a tone that it was scarcely possible to trace the rudiments of its future song in those early attempts. As it gained strength and confidence, however, its native notes became very apparent, and they continued to improve in tone till the termination of July, when it commenced moulting. . . . By the beginning of October . . . it began to execute its song in a manner calculated to remove every doubt as to its being that of the redbreast, had any such previously existed."

Mr. Long lays great stress on the manner in which parents inculcate into their young fear of enemies. Fear, he asserts, is not instinctive ; young creatures, if found before they have been taught to fear, are not alarmed at the sight of man. I admit that very young creatures are not afraid of foes, and that, later, they do display fear, but I assert that this change is not the result of teaching, that it is the mere development of an inborn instinct which does not show itself until the young are some days old, because there is no necessity for it in the earliest stages of the existence of a young bird.

Some months ago one of my *chaprassis* brought me a couple of baby red-vented bulbuls which had fallen out of a nest. They were unable to feed themselves, and were probably less than a week old. One met with an early death, and the survivor was kept in a cage. One day, while I was writing in my study, this young bulbul began scolding in a way that all bulbuls do when alarmed. On looking round, I discovered that a *chaprassi* had silently entered the room with a shikra on his wrist. The shikra is a kind of sparrow-hawk, common in India. That particular individual was being trained to fly at quail. It had never before been brought to my bungalow, nor is it likely that the captive bulbul, whose cage was placed in a small, enclosed verandah, had ever set eyes upon a shikra. It had left the nest before it was of an age at which it could learn anything from its parents. Its display of fear and its alarm-call were purely instinctive. Its inherited memory must have caused it to behave as it did. Speaking figuratively, its ancestors learned by experience that the shikra is a dangerous bird—a bird to be feared—and this experience has been inherited. To express the matter in more exact language, this inherited fear of the shikra is the product of natural selection. For generations those bulbuls who did not fear and avoid the shikra fell victims to it, while the more cautious ones survived and their descendants inherited this characteristic.

Of all the arts practised by birds none is so wonderful as that of nest-building. If it can be demonstrated (as I believe it can) that this art is innate in a bird, then there is no difficulty in believing that all the other arts practised by the feathered folk are innate.

Michelet boldly asserts that a bird has to learn how to build a nest precisely as a schoolboy has to learn arithmetic or algebra. By way of proof, he quotes the case of his canary—Jonquille. " It must be stated at the outset," he writes, " that Jonquille was born in a cage, and had not seen how nests were made. As soon as I saw her disturbed, and became aware of her approaching maternity, I frequently opened her door and allowed her freedom to collect in the room the materials of the bed the little one would stand in need of. She gathered them up, indeed, but without knowing how to employ them. She put them together and stored them in a corner of the cage. . . . I gave her the nest ready made—at least, the little basket that forms the framework of the walls of the structure. Then she made the mattress and felted the interior coating, but in a very indifferent manner."

Michelet construes these facts as proof that the art of nest-building is not innate in birds, but has to be learned. As a matter of fact they prove exactly the opposite. The Frenchman's reasoning is typical of that of those persons who make their facts fit in with their theories. Michelet is blinded by his preconceived notions. He is unable to see things which should be apparent to all. If the art of nest-building is not innate, why did the canary fly about the room collecting the necessary materials and heap them in a corner of the cage? That she did not go so far as to build a nest is easily explained by the fact that she was not given a suitable site for it, that the necessary foundation of branches was not provided for her. As well might one say that a bricklayer did not know his trade because he failed to build a wall on the surface of the sea. When given the framework, Michelet's untaught canary lost no time in lining it. The alleged act that the lining was not well done may be explained in many ways. Michelet may have imagined this, or the materials provided may not have been altogether suitable ; moreover, Jonquille must have worked in haste, as the framework was presumably not given until the bird had collected all the material. Again,

the nest was the first that that particular canary had
built. Birds, like human beings, learn to profit by
experience. Nest-building is an instinctive art, but
intelligence may step in and aid blind instinct.

In this connection it is necessary to bear in mind
that the nest is completed long before the young birds
come out of the egg ; that they leave, or are driven
away from, the parents before the next nest-building
season. If young birds are taught nest-building, who
teaches them ?

Proof of the instinctiveness of nest-building might
be multiplied indefinitely. There are on record scores of
instances of birds selecting impossible sites for their
nests ; these are cases of instinct that has gone astray.
Again, the persistent way in which martins will rebuild,
or attempt to rebuild, nests that are destroyed, shows to
what an extent nest-construction is a matter of instinct.
One more concrete piece of evidence must suffice.
My friend, Captain Perreau, has, among other birds
in his aviary at Bukloh, in the Himalayas, some grey-
headed love-birds. This species has the peculiar habit
of lining the nest with strips of bark, which the hen
carries up to the nest amongst the feathers of the back.
Captain Perreau started with two cock love-birds and
one hen, and this last had the peculiarity of not
carrying up the lining to her nest in the orthodox way ;
nevertheless, her daughter, when she took unto herself
a husband, used to carry up bark and grass to her nest
in the orthodox manner. " Why did this hen do this ? "
Captain Perreau asks. " Her mother could not have
taught her. I have no other true love-birds ; and my

blue-crowned hanging parrakeets, or rather the hens, certainly do carry up to the nesting-hole bark, etc., but they carry it, not in the back, but tucked in between the feathers of the neck and breast." This neat method of conveying material to the nest is, therefore, certainly an instinctive act, as is almost every other operation connected with nest-building.

To sum up. The parental teaching forms a far less important factor in the education of birds than many naturalists have been led by careless observation to believe. Birds may be said to be born educated in the sense that poets are born, not made. In each case education puts on the finishing touches to the handiwork of nature.

XLIV

BIRDS AT SUNSET

IT is refreshing to watch the birds at the sunset
hour. The fowls of the air are then full to
overflowing of healthy activity.

In the garden the magpie-robin (*Copsychus
saularis*), daintily clothed in black and white, vigorously pours forth his joyous song from some leafy
bough. From the thicket issue the sharp notes of the
tailor-bird (*Orthotomus sutorius*), the noisy chatter of
the seven sisters (*Crateropus*), and the tinkling melody
of the bulbul.

The king crows (*Dicrurus ater*) are alternately
catching insects on the wing and giving vent to their
superfluous energy in the form of cheerful notes. Upon
the lawn the perky, neatly-built mynas are chasing
grasshoppers with relentless activity ; nimble wagtails
are accounting for numbers of the smaller insects, while
the showy-crested hoopoes are eagerly extracting
grubs and other good things from the earth by means
of their long forceps-like bill. All, especially the
hoopoes, have the air of birds racing against time.
On that part of the lawn which the *malli* is flooding

to preserve its greenness the crows are thoroughly enjoying their evening bath.

On the sandy path is a company of green bee-eaters (*Merops viridis*) engaged in dust-bath operations.

Overhead the swifts—our little land albatrosses—are dashing hither and thither at full speed, revelling in the abundant insect life called forth by the fading light, and making the welkin ring with their " shivering screams." Flying along with the swifts are some sand-martins (*Cotile sinensis*), easily distinguishable by their slower and more laboured motion.

High above the sphere of action of the swifts and martins are numbers of kites and vultures, sailing in circles on their quest for the wherewithal to satisfy their insatiable appetite.

As the darkness begins to gather these birds, one and all, put more energy into their movements. Each seems to be aware of the rapid approach of the night when work must cease, and each appears fully determined not to lose a moment of the precious daylight.

While the sun is still well above the horizon great flocks of mynas sweep swiftly overhead towards the dense clump of bamboo bushes in which they will spend the night. They are joined by other species of starling. Before settling among the bamboos they perch in trees hard by, and make a joyful noise; every now and then some of the throng take to their wings and perform, like trained soldiers, a series of rapid evolutions. When at length the gloom compels them reluctantly to desist from their vigorous exercise, and to disappear into the bamboo clump, they give out energy in the form of loud clamour.

From the grove of tall trees yonder, where thousands of crows will spend the hours of darkness, an even greater noise issues. Some twenty minutes before the sun dips below the horizon the advance guard of the corvi arrives ; then, for the succeeding quarter-hour, continuous streams of crows come pouring in from east and west, from north and south.

Meanwhile the sparrows have been foregathering in their hundreds in the low shrubs that fringe the edge of the garden. And what a dissonance issues from those bushes !

Truly phenomenal is the activity of the birds at eventide. It is especially marked in India, where during the middle of the day the sun nearly always shines fierce and hot, so that the birds are glad to enjoy a siesta in the grateful shade. From this they emerge like giants refreshed.

This liveliness of the feathered folk at sunset is no small matter. It is one of the most pleasing facts of natural history. It shows how immensely birds enjoy life. It proves how healthy, how full of energy they are, how they, to speak figuratively, live within their incomes.

Contrast such scenes as those described above with what may be seen in the City of London at six o'clock on a weekday. A multitude of pale, anxious, worn-looking men, and thin, tired, haggard women pursue with listless steps their homeward way. Compared with the lot of these, how happy is that of the birds.

Birds are, like children, loath to go to bed. They feel no weariness, and so great is their enjoyment of life, that they are almost sorry when the sun disappears for a little.

Jules Michelet, than whom no more wrong-headed naturalist ever lived, declares that birds dread the night. "Heavy," he writes, "for all creatures is the gloom of evening. . . . Night is equally terrible for the birds. . . . What monsters it conceals, what frightful chances for the bird lurk in its obscurity. Its nocturnal foes have this characteristic in common—their approach is noiseless. The screech-owl flies with a silent wing, as if wrapped in tow. The weasel insinuates its long body into the nest without disturbing a leaf. The eager polecat, athirst for the warm life-blood, is so rapid that in a moment it bleeds both parents and progeny, and slaughters a whole family.

"It seems that the bird, when it has little ones, enjoys a second sight for these dangers. It has to protect a family far more feeble and more helpless than that of the quadruped, whose young can walk as soon as born. But how protect them ? It can do nothing but remain at its post and die ; it cannot fly away, for its love has broken its wings. All night the narrow entry of the nest is guarded by the father, who sinks with fatigue, and opposes danger with feeble beak and shaking head. What will this avail if the enormous jaw of the serpent suddenly appears, or the horrible eye of the bird of death, immeasurably enlarged by fear ? "

Greater nonsense than this was never penned outside a political pamphlet. Birds do not, as Michelet seems to imagine, go to sleep quaking with terror. They know not the meaning of the word death, nor have they any superstitious fears of ghosts and goblins.

Birds with young sleep the sleep of a man without a single care.

At other times birds do not roost in solitude, but gather together in great companies, the members of which are as jolly as the young folks at a supper party after the theatre. The happiness of the fowls of the air at the sunset hour is almost riotous.

Darkness, however, exercises a soothing influence over them. A feeling of sleepiness steals over them, and they then doubtless experience the luxurious sensation of tiredness which we human beings feel after a day spent in the open air ; for, although they know it not, their muscles are tired as the result of the activity of the day.

Their sweet slumbers completely refresh them. Before dawn they are awake again, and are up and about waiting for it to grow sufficiently light to enable them to resume the interrupted pleasures of the previous day.

S

WITH the exception of "The Education of Young Birds," which came out in *The Albany Review,* the chapters which compose this book appeared in one or other of the following Indian periodicals: *The Madras Mail, The Times of India, The Indian Daily Telegraph, The Indian Field, The Indian Forester.*

The author begs to tender his thanks to the several editors for permission to reproduce this collection of essays.

GLOSSARY

Bandobast. Arrangement.

Bhimraj. The racket-tailed drongo (*Dissemurus paradiseus*).

Chabutra. A masonry platform, erected outside the bungalow in the compound on which people sit in the evenings during the hot weather.

Chamar. The name of a low caste of Indians who skin animals and tan their skin.

Chaprassi. Lit., a badge-wearer. A servant who runs messages.

Chik. A number of thin pieces of bamboo strung together to form a curtain. *Chiks* are usually hung in front of doors and windows in India with the object of keeping out insects, but not air.

Chota hazri. Early morning tea.

Dak bungalow. Government rest-house.

Jhil. A lake or any natural depression which is filled with rain water all the year round or only at certain seasons.

Kankar. Lumps of limestone with which roads are metalled in Northern India.

Koi Hai. Lit., Is anyone there ? The expression used in India to summon a servant, bells being non-existent in that country.

Lathi. A club or long stick often studded with nails to make it a more formidable weapon.

Madar plant. Calotropis gigantea.

Mohwa. Bassia latifolia.

Murghi. A fowl or chicken.

Nullah. A ravine.

Ryot. A cultivator or small farmer.

Sahib. Sir, or a gentleman. A term used to denote a European.

Sath Bhai. Any of the various species of Crateropus babblers.

Shikari. One who goes out shooting or hunting.

Swadeshi. A jingoistic term meaning Indian.

Terai. Lit., moist land. A low-lying tract of land running along the foot of the Himalayas.

Tope. A grove of trees.

Topi. A sun-helmet.

INDEX

INDEX

www.ingramcontent.com/pod-product-compliance
Lightning Source LLC
Chambersburg PA
CBHW020527270326
41927CB00006B/480